Liquid Piston
Stirling Engines

Liquid Piston
Stirling Engines

C. D. West
Oak Ridge National Laboratory

VNR VAN NOSTRAND REINHOLD COMPANY
NEW YORK   CINCINNATI   TORONTO   LONDON   MELBOURNE

Published by Van Nostrand Reinhold Company Inc.
135 West 50th Street, New York, N.Y. 10020

Van Nostrand Reinhold Publishing
1410 Birchmount Road
Scarborough, Ontario M1P 2E7, Canada

Van Nostrand Reinhold
480 Latrobe Street
Melbourne, Victoria 3000, Australia

Van Nostrand Reinhold Company Limited
Molly Millars Lane
Wokingham, Berkshire, England

15 14 13 12 11 10 9 8 7 6 5 4 3 2 1

Library of Congress Cataloging in Publication Data

West, C. D.
    Liquid piston Stirling engines.

    Includes indexes.
    1. Stirling engines.   I. Title.
TJ765.W47 1982   621.4   82-8411
ISBN 0-442-29237-6   AACR2

Preface

This book is concerned with a relatively new and undeveloped type of Stirling machine—the liquid piston engine. The book contains some new material as well as a synoptic description of most of the known published work in the field. No doubt it will become incomplete even before publication as new ideas are proposed and published. There are known to be new proposals for vapor-phase machines (Cooke-Yarborough at the Harwell Atomic Energy Research Establishment) and for compounded machines (Ted Finkelstein and others); perhaps we shall soon see details of these and other ideas to add to the material already discussed in this book.

It was a pleasure and a surprise when Professor Joe Walker (who has quite literally written the book—in fact, three books—on modern Stirling machines) suggested that I join him in preparing a book on free piston engines that would complete and round off his sequence of works on regenerative machines. In the event, the timetable and content have led us to decide on dividing the field into two separate books: Walker will soon finish the other one—*Free Piston Stirling Engines*. I am very grateful for his encouragement and for his knowledgeable advice on all aspects of the project: I am especially pleased that it has given us the chance to become better acquainted.

Our hope is that these works will be useful in stimulating further developments and the interest of fresh minds in new ideas and applications for these machines.

The figures and drawings were prepared by Burt Unterburger and his assistants at the University of Calgary; and the final draft was also typed there by Karen Undseth. Their efforts in coping with all this work are much appreciated.

My wife Suzanne had the appalling task of quickly preparing the first typed draft, on old-fashioned machinery, from my handwritten notes. I owe her the warmest thanks for that and for the generous way she and my son Peter accepted long hours of family separation during the weekends and vacations spent writing these chapters.

C. D. West
Oliver Springs
Tennessee

Contents

Chapter 1
Liquid Piston Engines

The earliest practical heat engine, a drainage pump patented in 1698 by Thomas Savery, was a liquid piston machine (Sandfort, 1964). Hand-operated valves admitted steam from a boiler into a separate vessel. The steam valve was closed, and the vessel was then cooled by water flowing over its outside surface. This condensed some of the steam, creating a partial vacuum which acted directly on the water to be pumped and drew some of it up into the vessel (Figure 1.1). By further manipulation of valves, steam was again admitted into the vessel, forcing water out and into a discharge pipe. Subsequent developments of the steam engine by Newcomen, Watt, and others led to improvements in output, efficiency, and flexibility; but Savery was probably the first man to make a commercial business out of engines.

Another liquid piston, patented by Piot in the 1890s and in a slightly different form by McHugh in 1926, is the well-known "putt-putt" toy (Figure 1.2). Recent developments of this machine by Payne and his colleagues (Payne, Brown, and Brown, 1979) have led to designs that can by no means be regarded as toys. Their work also illustrates clearly that the simplest of devices tend to involve some very subtle physics.

An experimental flash boiling liquid piston engine has been described by Murphy (1979). One form of this engine is shown in Figure 1.3. Both flame-heated and solar machines were tested. The efficiency based on the indicated power (i.e., not including losses) using Freon as the working fluid was just over 1 percent.

Internal-combustion liquid piston engines have also been built and sold. The first of these was probably the Humphrey pump, first described in 1909. This is a more-or-less normal internal-combustion engine, using either a two- or four-stroke cycle (Ewing, 1926) in which the bottom of the cylinder is formed by a liquid column instead of the conventional solid piston (Figure 1.4). Inlet and outlet valves, and an

1

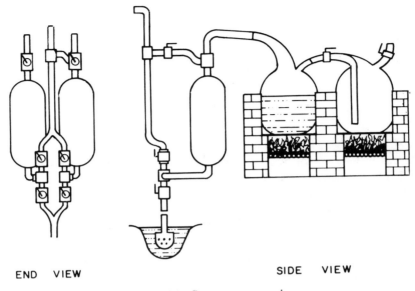

END VIEW   SIDE VIEW

Figure 1.1. Savery steam engine.

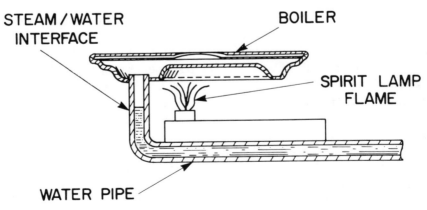

Figure 1.2. McHugh water pulsejet. (After Payne, Brown, and Brown, 1979.)

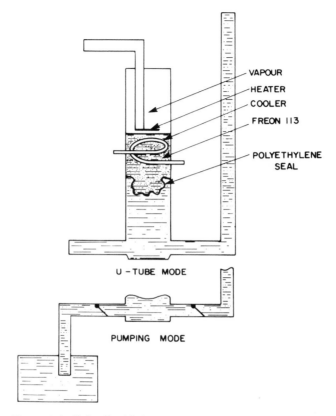

Figure 1.3. Solar liquid piston pump. (After Murphy, 1979.)

ignition system, are necessary; and of course a high-grade fuel—usually petroleum or gas—must be used.

Gongwer (1950 and 1960), working at the Aerojet General Corporation, described a water pulsejet in which the expanding gas from fuel ignited in a separate chamber forces water out of a pipe in a propulsion jet (Figure 1.5).

Morash and Marshall (1974) describe an engine patented by Roesel (1970) in which heat is transferred by a hot or cold liquid sprayed into a closed chamber containing the working gas. The working gas is expanded or compressed by pistons. Although conventional pistons

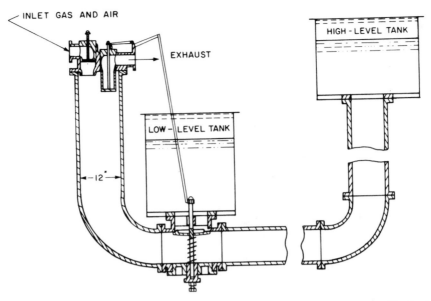

Figure 1.4. Humphrey liquid piston internal combustion pump. (After Sir J. A. Ewing, 1926.)

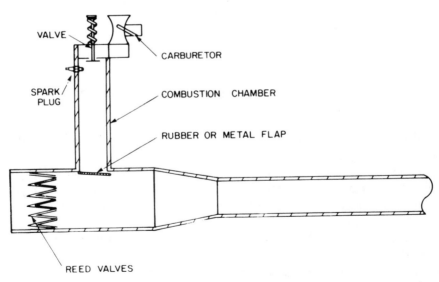

Figure 1.5. Aerojet internal-combustion water pulsejet for underwater operation. (After Payne, Brown, and Brown, 1979.)

could, of course, be used, the engine was normally designed to use a liquid piston (Figure 1.6). By varying the timing and duration of the heat-transfer spray, different thermodynamic cycles could be used. It is hoped that further work will be done on this very interesting external-combustion-engine concept.

The "thermal bounce" engine described by Lloyd (1975) consists in one form of a U tube of mercury ending in two bulbs, each containing an electric filament (Figure 1.7). When the filaments are heated, a vigorous oscillation of the mercury is observed. The mode of operation of the engine is uncertain and little power can be extracted from the small models used. It is hoped that more work will be reported on this idea, for almost nothing seems to be known about it.

Malone (1931) designed and built an engine in which the working fluid itself is a liquid, undergoing a closed regenerative cycle somewhat similar to a Stirling cycle (Walker, 1980; Allen, Knight, Paulson, and Wheatley, 1980). In Malone's engine, the forces and displacements

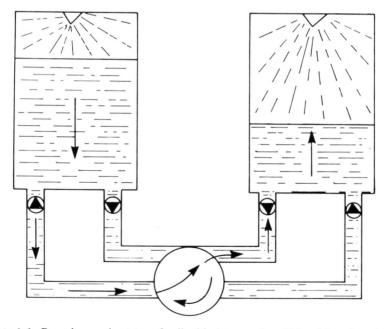

Figure 1.6. Roesel spray heat-transfer liquid piston engine. (After Morash and Marshall, 1974.)

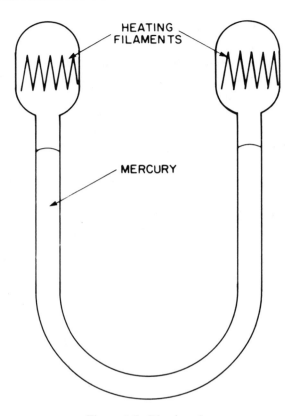

Figure 1.7. Lloyd engine.

were originated in the liquid but ultimately transmitted by solid pistons. Although generations of physicists and engineers have been taught to write of working "fluids," this is actually one of the few practical cycles for which the more restrictive term of "working gas" would not be accurate over any part of the cycle.

In the course of their long and extensive Stirling machine research program, the Philips company patented an engine configuration in which solid pistons were coupled together hydraulically through a liquid medium (Dros, 1965). Later, workers at Reading University (Dunn, Rice, and Thring, 1975) designed a Stirling engine with no solid power piston. In its place was a rubber bag filled with hydraulic fluid. It was also suggested that a hydraulic system could be used to drive the solid displacer system (Figure 1.8).

The major advantage of liquid pistons is clearly seen in the putt-putt steam engine and the Humphrey internal-combustion engine: it is simplicity. Liquid pistons do not require accurately dimensioned cylinders and they permit great flexibility in mechanical design with relatively simple construction. If the liquid is water, then an engine and pump can be made as a single system.

These are also the characteristics of liquid piston Stirling engines (West, 1970a), although it should be remarked immediately that many practical examples involve a two-phase working fluid and are thus not true Stirling machines.

The basic principle of the liquid piston Stirling engine, known as the Fluidyne, is shown in Figure 1.9. In this case, both the displacer and output pistons are liquid. As the displacer liquid column oscillates in its U tube, the gas above the liquid surface is transferred back and forth between the hot and cold spaces. The resulting pressure variations act

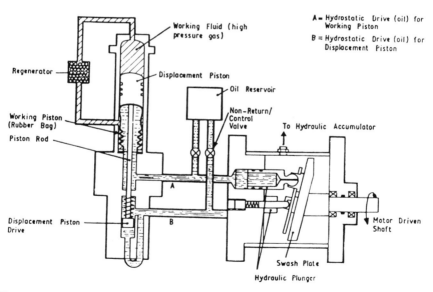

Figure 1.8. Enclosed-liquid power piston engine. (After Dunn, Rice, and Thring, 1975.)

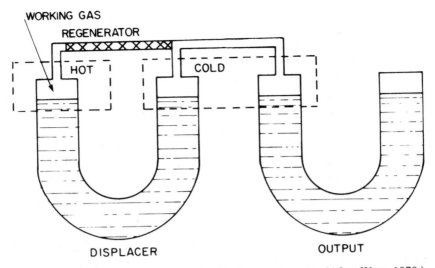

Figure 1.9. Basic Fluidyne without feedback arrangements. (After West, 1970.)

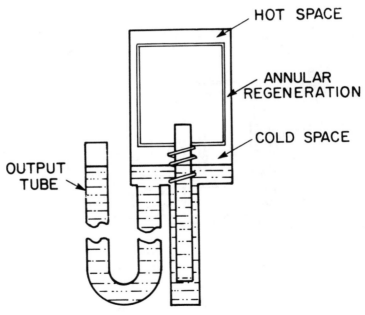

Figure 1.10. Solid displacer, liquid output piston engine. (After West, Cooke-Yarborough, and Geisow, 1970.)

upon the liquid in the output column, causing it to move also. In Chapter 2, we shall see several ways of maintaining the displacer in oscillation and several different configurations of this basic engine.

The use of a liquid-output piston avoids the need for a sliding mechanical seal, which has been a continuing problem for crankshaft Stirling engines. A free-displacer-piston machine does not require such a seal on the displacer; and one form of the Fluidyne takes advantage of this by using a liquid-output piston but a "solid" (probably hollow) displacer (see Figure 1.10), (West, Cooke-Yarborough, and Geisow, 1970; Martini, Hauser, and Martini, 1977; Goldberg and Rallis, 1979). This kind of machine has the advantage that the hot end temperature is not limited by the properties of the liquid used.

Most interest so far has centered around the use of Fluidyne machines to pump water, and several groups around the world are or have been working toward this end. Where possible, their work is cited in the references. The Metal Box Company of India Ltd. maintained an effort on the development of Fluidyne irrigation pumps after the end of their joint research program with the Harwell Laboratory in England in 1977, but this effort seems to have been terminated in early 1981. The Metal Box machines have a much higher performance and better efficiency than any described in the open literature, but the company has, for commercial reasons, released little information about their present designs, although details are now available about the earlier experimental models (West and Pandey, 1981).

Like any other Stirling machine, the liquid piston engine can also be operated as a refrigerator or heat pump; several workers have proposed exploiting this. Walker has suggested that a liquid helium cooler—to service superconducting computers, for example—could be made in which the liquid helium would be used in a Fluidyne machine driven by gas-pressure variations from outside the cryogenic region.

The following chapters describe the principles and design characteristics of the liquid piston Stirling engine, or Fluidyne. Because so many of the published descriptions refer to machines in which evaporation is permitted, and in which the vapor pressure plays an important role, a discussion of this effect is also given.

Chapter 2
Basic Operation of the Fluidyne

The basic principle of the Stirling engine is a simple one: it relies only on the fact that when a gas is heated, it tends to expand or, if confined, to rise in pressure. Consider first the arrangement shown in Figure 2.1. A single piston is placed within a cylinder whose opposite ends are interconnected by a tube. A gauge registers the pressure of the gas in the cylinder, which is heated at one end and cooled at the other.

At the top of Figure 2.1 the piston is shown in the center of the cylinder; half the gas is at the hot end, half of it is at the cold end, and the pressure gauge reads normal.

In the center diagram, the piston has moved to the cold end of the cylinder and this moves, or displaces, some gas from that end through the tube to the hot end. The average temperature of the gas is therefore higher, and its pressure goes up, as indicated by the gauge.

The lower diagram shows that the piston has moved to the hot end, displacing the gas through the tube and into the cold end. The average temperature of the gas is now lower, and its pressure falls, as observed on the pressure gauge. The only function of the piston—the displacer— is to move the gas from the cold to the hot end of the cylinder and back again.

Figure 2.2 shows how the displacer pressure changes can be used to drive another piston and do work. This second cylinder and piston are often called the expansion cylinder and the power piston. When the gas pressure is high (displacer piston at the cold end, gas at the hot end), the power piston is allowed to move toward the open end of the expansion cylinder and in doing so it can perform work—moving a pump handle to raise water, for example, or turning a crankshaft.

When the gas pressure is low (displacer piston at the hot end, gas at the cold end), the power piston is returned to its original position. Some work is needed to do this, of course, but it is less than the work which

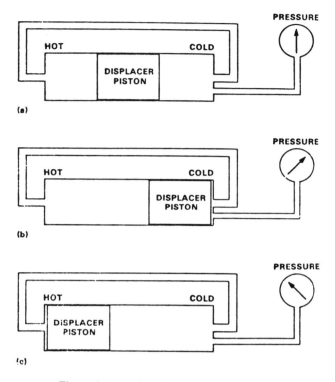

Figure 2.1. Action of the displacer piston.

was made available when the power piston moved out, because the force on this piston is now less, owing to the reduced gas pressure. Thus, over the complete cycle, more energy can be taken out of the power piston than needs to be put in; and this excess of energy can be used to turn a shaft or to operate a pump or to perform any of the other duties expected of an engine.

A more subtle point concerns heat regeneration. In most practical machines the tube connecting the cylinders is filled with what is known as the regenerator; this may be a series of wire-mesh baffles or a set of narrow passages through which the gas must flow on its journey between the hot and cold regions (Figure 2.3).

The function of the regenerator may best be explained by first considering what would happen in its absence. As the displacer piston is

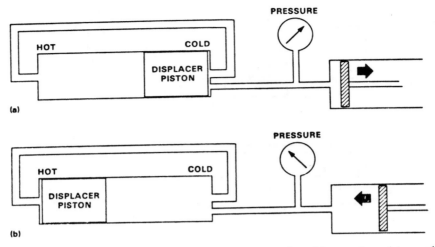

Figure 2.2. The displacer pressure changes can be used to drive another piston and do work.

moved from right to left, hot gas would flow through the connecting tube and into the cold end of the cylinder; when it arrived there, it would be cooled down and the heat extracted from it during this cooling process would have to be carried away by the cooling water or air which was being used to keep the right-hand end of the displacer cylinder at a low temperature. This heat would therefore be wasted—and, of course, wasting heat reduces the efficiency of the engine.

With the regenerator present, however, there is a steady fall in temperature along the regenerator, from left to right, as the gas gives up

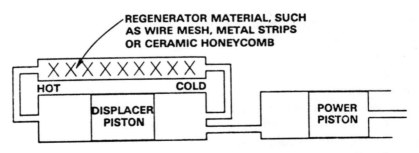

Figure 2.3. The regenerator acts to increase efficiency but not necessarily to increase the power.

heat to the regenerator material. By the time the gas emerges into the cold end of the cylinder, therefore, it has already been cooled and no extra heat has to be carried away by the coolant.

When the cold gas flows back to the hot cylinder, its temperature gradually rises as it picks up the heat left in the regenerator during its journey to the cold end. This heat is thus not wasted; the regenerator operates as a kind of heat store, and the efficiency of the engine is therefore increased.

THE LIQUID PISTON FLUIDYNE ENGINE

With this description of the Stirling engine, the operation of the basic Fluidyne, shown in Figure 1.9, is easy to explain. The left-hand U tube (which has one end heated and the other end cold) functions as the displacer; and the right-hand tube, which has one end open to the atmosphere, works as the output, or power, piston. A conventional Stirling engine with this configuration is called a gamma configuration machine. Suppose that we set the water in the displacer oscillating from one limb of the U tube into the other limb and back. Top dead center in the cold end will correspond to bottom dead center in the hot end; this situation is illustrated in the left-hand part of Figure 2.4, in which most of the air trapped above the water in the displacer is in the hot left-hand limb. Most of the air is therefore hot, so its pressure will rise, which tends to

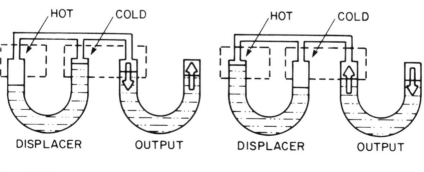

Figure 2.4. Basic operation of the Fluidyne.

force the water in the output tube to move from right to left, as the arrow indicates.

Half a period later the displacer water will have swung back into the other limb, so that the cold surface is at bottom dead center: this is the situation in the right-hand sketch of Figure 2.4. Most of the air is now in the cold side of the machine, so its pressure will fall, pulling the water in the output column back from left to right.

TUNING OF LIQUID COLUMNS

As in any oscillating system, the maximum amplitude of movement in the output column will be achieved if the frequency of the pressure variation, i.e., of the driving force upon it, is about equal to the natural or resonant frequency of the water oscillating in the output column. These pressure variations are due to the oscillations of the displacer water, so it follows that for maximum movement we must make the two natural frequencies equal.

What happens if the water-column length in the output tube is not adjusted to give it the same natural frequency as the displacer? If the column is too long, the mass of water in it will be so great that the pressure change will be unable to move it very far: there will be almost no change in volume of the working gas during the cycle, and hence almost no work will be done. On the other hand, if the water column is too short, it will move so easily that the pressure in the engine will be unable to build up significantly before the column moves to its full extent, and again almost no work will be done. The length of the output U tube must therefore be "tuned" to suit the operating frequency of the engine.

As long as the water (or other liquid) in the displacer can be kept oscillating back and forth in its U tube, the water in the output tube will also move back and forth, taking the energy that it needs to do this from the changing air pressure in the machine. This changing pressure does not, however, have any effect upon the displacer, for it acts equally on both ends of the displacer column. If the displacer were left to itself, therefore, the oscillations in the displacer tube would eventually die away because of viscous friction and other losses; with them would die the displacer action and the movement of the output column.

ROCKING BEAM FEEDBACK ENGINES

Some means must be found to keep the displacer in motion. Several ways of doing this are known now but the earliest method was called "rocking beam feedback" (West, 1970*a*, 1971). The whole machine is mounted on a pivot, as shown in Figure 2.5, although the pivot could be replaced by a flexure or spring mounting that would not be subject to wear or static friction. As the liquid in the lower (output) tube moves back and forth, its shifting weight causes the whole machine to rock like a seesaw and this movement keeps the displacer liquid in motion. In effect, part of the energy of the output column motion is fed back, by means of a rocking motion, to the displacer column to keep it in motion also. A spring may be used to give some extra restoring force and adjust the frequency of the rocking motion to match, more or less, the other frequencies in the system. This arrangement was used, in conjunction with a simple valveless pump mounted on the same pivot (Figure 2.6), to demonstrate that the Fluidyne could do external work by pumping water—albeit on a very small scale at that time.

This type of rocking beam feedback depends on the shifting weight

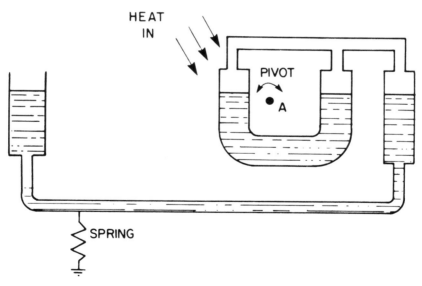

Figure 2.5. Rocking beam feedback for a displacer Fluidyne engine.

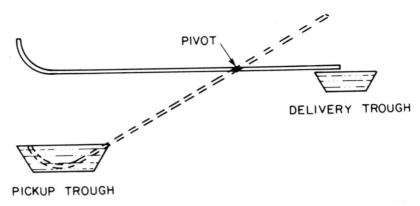

PICKUP TROUGH

Figure 2.6. Low head pump for a rocking beam machine.

of liquid to maintain the rocking motion. Another way to achieve the same result is given in a Fluidyne patent (West, 1974*a*) and illustrated in Figure 2.7. In this case the pressure variations inside the machine act on a bellows or other flexible coupler to provide the movement necessary to keep the beam rocking.

It seems that by suitable choice of the bellows diameter and position, and of the moment of inertia of the rocking components, the tuning column could be dispensed with and pumping power taken either from the gas-pressure variations or mechanically from the rocking motion: however, this possibility has apparently never been fully worked out, nor tried experimentally.

The rocking beam machine has many potential attractions, including the fact that the displacer and output columns can be separate and could contain different liquids—for example, water in the output column, for cheapness, but oil in the displacer so that higher temperatures may be used at the hot end without boiling or excessive evaporation. Also, the mechanical motion could be readily converted to rotating motion with a crankshaft, and be available to drive other machines, while retaining most of the simplicity of construction. These advantages have not yet been fully explored and almost all the theoretical work on Fluidynes reported so far has concentrated on feedback systems, to be described below, that do not require any moving parts. There is an open field for anyone who seeks to explore further the rocking beam type of

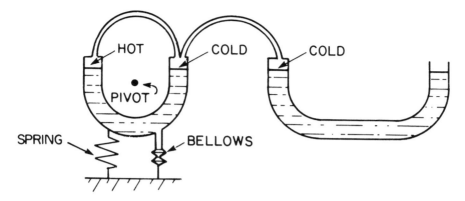

Figure 2.7. Rocking beam engine driven by pressure variations in the working fluid.

machine and to understand fully its practical advantages and disadvantages.

One obvious disadvantage of the rocking beam machines described so far is that, like all gamma-form engines, they have three separate cylinders. Although this does, as indicated, allow a free choice of displacer liquid, it also increases the minimum volume of the working gas, which is the volume when the output piston is at top dead center. In fact, even when the output cylinder volume is zero, i.e., at top dead center, the gas volume is still equal to the stroke of the displacer plus the volume of the regenerator. This naturally reduces the compression ratio and hence the specific power output, although the advantages already referred to may easily outweigh this consideration in specific cases.

A simple variation in the construction (Figure 2.8) reduces the number of cylinders to two, in the alpha configuration of a Stirling engine. The compression ratio can now be higher, but the possibility of using two different liquids not in contact with each other has been lost, unless some kind of separating membrane is used.

LIQUID FEEDBACK ENGINES

Most work so far has concentrated on Fluidynes without mechanical moving parts, because of their simplicity and freedom from mechanical

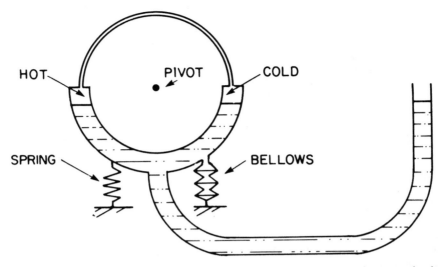

Figure 2.8. Pressure-driven feedback for a rocking beam machine in the merged cylinder or alpha configuration.

friction and wear. An example of this type of engine, in which feedback is provided directly by the motion of the liquid itself, is shown in Figure 2.9. Although this configuration, and others that are mathematically similar to it, were described in the original Fluidyne report and patent, it was not until much later that the basic theory of their operation was fully understood (Elrod, 1974; and Geisow, 1976). Recent work has considerably extended our theoretical knowledge of the dynamics of the liquid feedback system (Stammers, 1979).

To understand the operation of these machines, we first recall that to operate as an engine, the movement of the hot cylinder must lead the movement of the cold cylinder—that is, the hot piston must reach bottom dead center somewhat earlier than the cold piston. In the engine shown in Figure 2.9, the length of the hot column—measured from the free surface to the junction with the output tube—is less than the length of the cold column. Consequently, there is less mass of liquid in the hot side of the machine than in the cold, and when the pressure varies in the working fluid, the hot column responds more quickly than the cold, i.e., the hot-cylinder volume variation leads the cold.

From this argument (which is rather oversimplified), we should expect that the machine will run as an engine whether the end of the output tube points toward the hot end (as in Figure 2.9) or toward the cold end (as in Figure 2.10) or is at right angles to the displacer cylinder, as it is for some of the machines described in Chapter 4. On the other hand, if we transpose the hot and cold cylinders, by heating the end that is more remote from the output tube junction, the phasing will be wrong and the machine will not run. This is consistent with the author's experience gained in Fluidynes operating without evaporation, on a Stirling-like cycle. However, Goldberg (1979) observed that operation of a machine with evaporation could be achieved with the junction close to either the hot or the cold cylinder—but only if the end of the output line was directed toward whichever working space (hot or cold) was the closer. At present there is no explanation for these differences in observed behavior.

Elrod's theory (1974) describes the mathematical behavior of liquid feedback machines. His theory also includes a prediction for the temperature difference between the hot and cold ends that is necessary for the machine to break into oscillation and run as an engine. Interest-

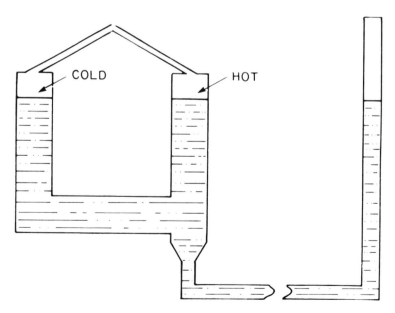

Figure 2.9. Liquid feedback machine (tuning line pointing toward hot end).

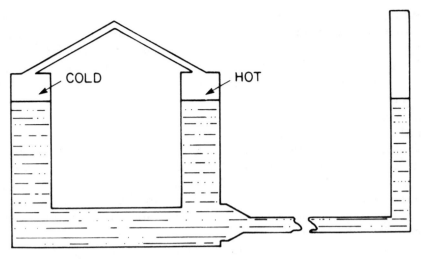

Figure 2.10. Liquid feedback machine (tuning line pointing toward cold end).

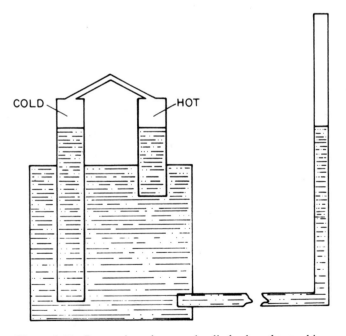

Figure 2.11. Reservoir and unequal cylinder lengths machine.

ingly, the theory shows that a finite, although quite small, temperature difference is needed even for an ideal machine with no losses whatsoever. In practice, an even larger temperature difference will be needed to overcome the losses due to viscous friction and imperfectly isothermal behavior of the cylinders (West, 1980) and indeed much larger differences have been noted experimentally [some results are given by Stammers (1979)].

Figure 2.11 illustrates a similar liquid feedback design in which the hot and cold columns, of different length, communicate with each other and with the output tube through a common reservoir. Figure 2.12 is a variant of this in which the hot- and cold-cylinder tubes are of equal length but the output tube is partially inserted into the hot cylinder to

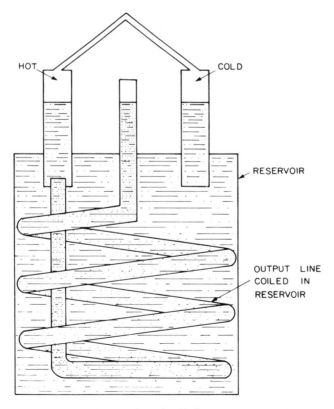

Figure 2.12. Reservoir and coiled output tube machine.

give an effective difference in the liquid-column lengths; as shown, the output tube can be coiled inside the reservoir, leading to a relatively compact and robust construction.

Figure 2.13 shows another form of liquid feedback machine in which the hot and cold columns are concentric cylinders (West, Geisow, and Pandey, 1976). This type of machine can conveniently incorporate one of several ways to package the tuning column into a smaller space—see for example Figure 2.14.

FREQUENCY LIMITATIONS

As we saw earlier, the operating frequency of these machines is determined primarily by the natural frequency of oscillation of the liquid

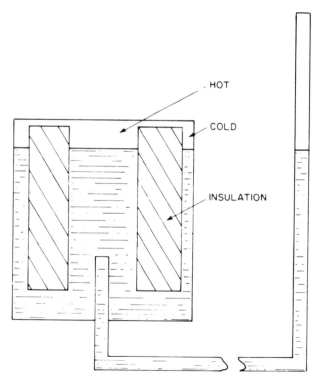

Figure 2.13. Concentric cylinder machine.

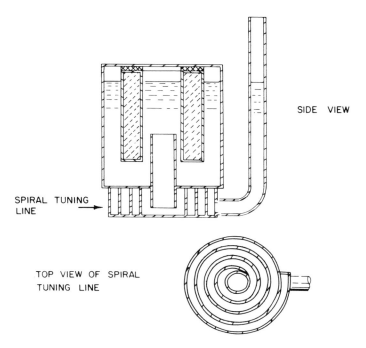

SIDE   VIEW

SPIRAL   TUNING
LINE

TOP VIEW OF SPIRAL
TUNING   LINE

Figure 2.14. Concentric cylinder machine with compact tuning line.

columns. The displacer is basically a simple U tube of liquid oscillating, like a pendulum, under the influence of gravity. This limits the range of frequencies that are practically achievable: this point is discussed in detail in Chapter 3, where it is shown that the range of frequency available with conveniently sized tubes is typically 0.4 to 1.3 Hz. This may be too restrictive for some applications; fortunately, there are other configurations that do not suffer the same limitations.

One of these is the rocking beam engine that has already been described. If the displacer U tube is rocked at a much higher frequency than the natural frequency of the liquid within it, there will be a strong tendency for the liquid to remain almost unmoving while the U tube oscillates past it. The effect is of a free piston engine with a very heavy piston and a lightweight cylinder. The result is to vary the hot- and cold-gas volumes at a frequency determined by the rocking beam and the output column instead of by the displacer column. In this way, a higher operating frequency could be used. A similar effect can be obtained by rocking the beam at a frequency much lower than the natural frequency of the displacer liquid, thus extending the range of operating frequency downward. As far as is known, neither method has been experimentally verified.

MULTICYLINDER FLUIDYNES

As long as the restoring force on the displacer liquid is provided only by gravity, its natural frequency of oscillation is determined by the effective length, taking into account changes in cross section, of the liquid column. The pressure variations induced by the displacer movement (or by any other mechanism) do not affect the displacer oscillation because they act equally on both the hot and the cold end of the column.

The output column, on the other hand, has another restoring force on it because movement of the output liquid also compresses or expands the working fluid and the resulting pressure changes react back on one end only of the column. This extra restoring force means that the output column must be longer than the displacer column, other things being equal, in order to have the same natural frequency. Conversely, if the output and displacer columns were identical, the output column would have a higher natural frequency. The effective compressibility of the gas depends on the temperature of the different working spaces; it is also dependent on the pressure and hence on the amplitude. Therefore we would expect some fairly small variation of natural frequency with operating conditions and amplitude.

By using a multicylinder configuration (West, 1974b), a system can be designed in which all liquid columns are subject to gas-pressure forces as well as to gravity (Figure 2.15). A similar arrangement was proposed independently by Finkelstein, and was the subject of some experimental measurements by the Chicago Bridge and Iron Company (Cutler and Hanke, 1979). Multicylinder free piston Stirling machines can also be configured as a heat-actuated heat pump (Cooke-Yarborough, 1975; Gerstmann and Freidman, 1977) and liquid piston versions of this heat pump are also possible (West, Geisow and Pandey, 1976; AMTI, 1982). The concentric cylinder arrangement can also be used for multicylinder machines (West, Geisow, and Pandey, 1976).

In a multicylinder machine, the forces acting on all the liquid col-

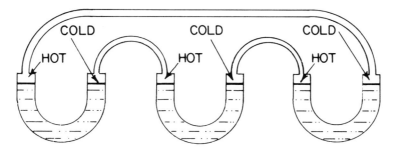

Figure 2.15. Multicylinder machine.

umns include both gravity and gas pressures. The latter are dependent on the volume of the working gas because a small gas volume will experience a greater pressure change, and hence exert a larger force, for a given volume change than will a large gas volume. Gravity is fixed, but the gas volume is a design parameter and can be varied, within other design constraints, to give a wider range of natural frequencies than is available from gravity alone. In a sealed system, such as the heat pump arrangement, the gas pressure can be increased, without much added complication, which will raise the natural frequency as well as increasing the specific output directly.

PUMPING CONFIGURATIONS

So far we have discussed several different forms of the liquid piston Stirling engine but said little about how the engine output power is to be extracted and used. Most interest in the Fluidyne to date has focused on its application for pumping, including irrigation pumping. Besides direct mechanical output from the rocking beam machines, there are three simple ways to use the Fluidyne output to pump water.

The first, known as series coupling (Figure 2.16), simply requires a T piece at the end of the output tube, and two nonreturn valves. On the inward stroke of the output liquid, when the gas pressure inside the engine is low, liquid is drawn in through the lower nonreturn valve. On the outward stroke, liquid is forced out through the upper valve. Most of the small working models of Fluidyne pumps that have been publicly demonstrated use this method.

In some circumstances, there are drawbacks to this arrangement. The

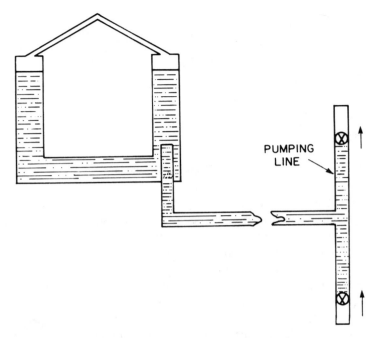

Figure 2.16. Pump in series with output column.

volume change within the engine is, of course, equal to the stroke in the output tube which, in turn, is equal to the amount of liquid pumped through the valves in each stroke. Above a certain pumping head, the work required to pump a particular volume of liquid becomes greater than the work done by the same volume change in the engine. Beyond this point, which will depend on the cylinder temperatures, the machine cannot pump at all. Furthermore, the presence in series with the output column of the nonreturn valves may upset the tuning of the system. Also, the valves are nonlinear, which may seriously affect the behavior of the output tube as a resonant, oscillating system. Nevertheless, for small engines, at least, the system works well.

By placing the pumping system closer to the displacer end of the output tube so that it is in parallel with the output column (Figure 2.17), some relief can be obtained from the drawbacks of the series arrangement. The volume of liquid moving in the output tube can be greater than the volume passing through the pump, so that there need not be the same close correlation between pumped volume and engine stroke as there is in the series arrangement. Similarly, nonlinearities in the flow through the pump will have a relatively smaller effect on the oscillation of the larger amount of liquid in the remainder of the output line. In this case, the output line does not do any work directly, except to overcome its own losses, but merely oscillates at a frequency tuned to that of the displacer, thus giving rise to a relatively large pressure variation, within the engine, for pumping. For this reason, the "output tube" in this case is usually referred to as the "tuning line" or "tuning column," recognizing that its main function is to have a large, resonant oscillation and not to provide a direct power output mechanism.

A simple variation on this arrangement disconnects the pump from the tuning column altogether (Figure 2.18) and drives it instead from

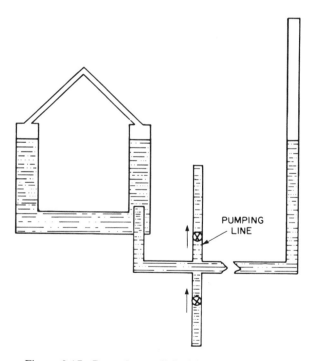

PUMPING
LINE

Figure 2.17. Pump in parallel with tuning column.

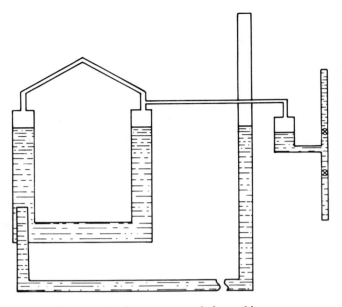

Figure 2.18. Pump gas-coupled to cold space.

the pressure variations in the working gas. When the pressure is low, liquid is drawn up through the lower nonreturn valve and into the pumping arm. When the pressure rises, this liquid is expelled through the upper valve. This configuration minimizes the direct interaction between the pump and the tuning line, although of course the amount of water pumped (i.e., the change in water level in the pumping arm) continues to affect the pressure variations in the machine. This method of connecting the pump is usually known as "gas coupling" although some people have referred to it as "soft coupling." It can, in principle, be applied to the rocking beam, liquid feedback, and multicylinder machines alike. It has the further advantage, first proposed by Cooke-Yarborough, that the pump and engine need not be at the same level. A more detailed discussion of gas coupling is given in the appendix.

Both series- and gas-coupled pumping systems have been made at Harwell using fluidic valves connected to liquid feedback Fluidynes. This results in a simple pump having no solid moving parts, and hence no solid wear or friction whatsoever.

EFFECTS OF EVAPORATION AND MEAN PRESSURE

At this point, it would be well to note that many of the experimental machines described in the open literature [exceptions are provided by Martini, Hauser, and Martini (1977); Goldberg and Rallis (1979); and West and Pandey (1981)] have permitted substantial evaporation to take place from the liquid in the hot cylinder. The most obvious effect of this evaporation is to increase the heat input required, since latent heat must be supplied. A second effect is to increase the power output. This is because the pressure variations are enhanced. For example, cooling a constant volume of air from 100 to 30°C could lower its pressure from 1 to 0.81 atm; by contrast, the saturated vapor pressure of water falls in the same temperature range from 1 atm to 0.04 atm, a five times larger change.

For a small machine, with many thermal, thermodynamic, and mechanical losses, the increased power availability due to evaporation is an almost unmixed blessing (Reader and Lewis, 1979a). However, the water system may be limited to an efficiency of 1 percent or less at these temperatures (Goldberg, Rallis, Bell, and Urieli, 1977; West, 1982b) and for a larger machine designed for high efficiency, there are advantages to suppressing evaporation. It is reported (Bell and Goldberg, 1976; Reader and Lewis, 1979b) that at hot-end temperatures of 370 K or below, it is the air component and not the vapor component of the working fluid that is dominant in the cycle, even when evaporation is not suppressed. However, it should be noted that the present author has observed that evaporation has a very marked effect at approximately this temperature in small engines with a displacement of about 10^{-6} m^3 (1 cm^3). All the evaporative cycle liquid piston engines— sometimes called "wet" Fluidynes—reported in the literature show a maximum efficiency of much less than 1 percent. By suppressing evaporation, at least 10 times greater efficiency can be achieved (West and Pandey, 1981; Pandey, 1981a), but a larger machine will be needed to make up for the lower specific power of the "dry" machine. A more detailed discussion of these effects is given in Chapter 8.

Another way to increase the specific output power of a machine, without incurring the large heat losses that accompany evaporation, is to increase the mean pressure of the working fluid. Pressurization is, of course, easier to achieve in a machine that does not require any outlet to the atmosphere (such as the multicylinder heat pump) than it is in the case of a pump. However, several ways of pressurizing a pumping Fluidyne have also been proposed (West, Geisow and Pandey, 1977). The operation of a pressurized machine has been experimentally verified at Harwell, but no quantitative measurements were made. It is expected that a pressurized machine would have a higher pumping-head capability and a higher efficiency than an atmospheric pressure engine, at the price of increased complexity.

In this chapter, we have discussed the basic operation of the Fluidyne, and some of the wide range of configurations that are known. Doubtless there are others, for the free-piston-engine concept is a broad one and the use of liquid components allows, in some cases, an added dimension of design freedom. In its present forms, the Fluidyne seems best suited to pumping applications, although other uses, including refrigeration and heat pumping, have been proposed. In later chapters, we discuss the design principles in more detail, and consider some of the thermodynamic, fluid flow, and thermal losses that are important in this kind of engine, and that set limits on its performance.

Chapter 3
Basic Design and Power Calculations

To get the maximum amplitude of oscillation in the liquid columns, the flow losses should be low and the frequency of operation should be close to the natural, or resonant, frequency of the columns themselves. Fortunately, these frequencies are fairly easy to calculate, especially if the losses are indeed kept small.

DISPLACER FREQUENCIES

To illustrate this, let us first consider the displacer alone: in this case, we can imagine both ends of the column to be open to the atmosphere. (Even if they are not actually open, the same gas pressure will always be acting, on both ends of the column, and its effect will cancel out) (see Figure 3.1).

To start the oscillation going, we raise the liquid level slightly in one arm of the tube—for example, by blowing down the other end. If the liquid surface rises by a distance x at one end, it must fall by the same amount at the other. One end of the liquid column now has more weight of liquid than the other, by an amount $2xA_d\rho$. The pressure arising from this is $2x\rho g$, where g is the acceleration due to gravity, and the resulting force is $2A_dx\rho g$. The mass of the liquid column is $A_d\rho L_D$, and so the acceleration induced by this force (in a direction to reduce x) is simply given by

$$A_d\rho L_D \ddot{x} = -2xA_d\rho g \qquad (3.1)$$

or

$$\ddot{x} = \frac{-2gx}{L_D}$$

This is the equation for undamped simple harmonic motion and the natural frequency, ω, is therefore given by

$$\omega = \sqrt{\frac{2g}{L_D}} \quad \text{rad/s} \qquad (3.2)$$

or
$$f = \frac{1}{2\pi} \sqrt{\frac{2g}{L_D}} \quad \text{Hz} \qquad (3.3)$$

In practice, a length less than about 0.3 m or more than about 3 m (1 to 10 ft) may be inconvenient: this puts the natural frequency in the range of 0.4 to 1.3 Hz. In seeking a preliminary feel for the operation of a machine with this kind of displacer, one need not go far wrong by assuming a frequency of about 0.65 Hz.

Now consider the more complex, but sometimes useful, case where the area of the tube connecting the two ends of the displacer is different from the area of the cylinders themselves (Figure 3.2). To make the analysis simple, without much loss of accuracy in practice, assume that the length of the enlarged sections is much less than the length of the connecting tube.

Once again, one surface of the liquid is raised by x, and the other therefore falls by the same amount. The extra height, $2x$, of one surface over the other, again gives rise to a pressure difference $2\rho gx$, which acts on the liquid column. Ignoring the relatively short and slow moving length of liquid in the enlarged section of the cylinders, we may consider

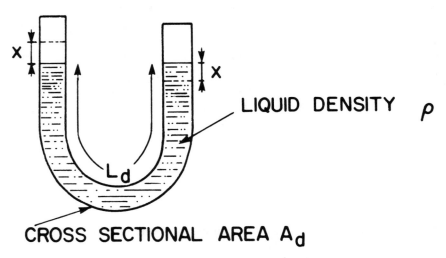

Figure 3.1. Simple displacer U tube.

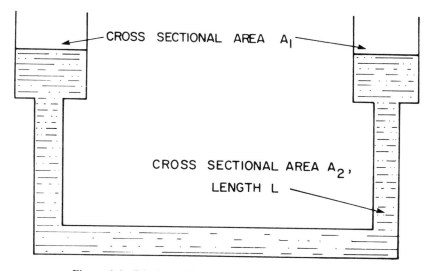

Figure 3.2. Displacer U tube with varying cross section.

only that the pressure $2\rho gx$ acts on the cross-sectional area A_2 of the connecting tube, which contains a mass of liquid ρLA_2. A movement x of the liquid surface gives rise to a movement $(A_1/A_2)x$ of the liquid in the connecting tube because of the difference in areas. The acceleration in the connecting tube is of course higher by the same ratio. Therefore,

$$\rho LA_2\left(\frac{A_1}{A_2}\right)\ddot{x} = -2\rho gxA_2 \tag{3.4}$$

$$\ddot{x} = \frac{-2g}{L}\frac{A_2}{A_1}x \tag{3.5}$$

The natural frequency is therefore

$$\omega = \sqrt{\frac{2g}{L}\frac{A_2}{A_1}} \quad \text{rad/s} \tag{3.6}$$

or

$$f = \frac{1}{2\pi}\sqrt{\frac{2g}{L}\frac{A_2}{A_1}} \quad \text{Hz} \tag{3.7}$$

By making the connecting tube much smaller than the cylinders ($A_2 \ll A_1$), the frequency can be lowered considerably: however, the liquid in

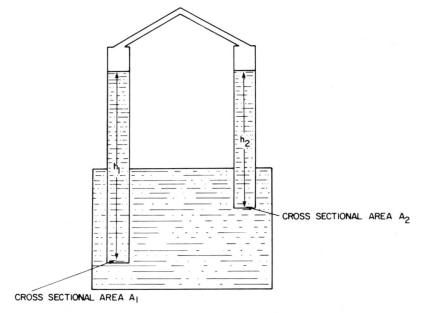

Figure 3.3. Reservoir displacer.

the connecting tube will be moving much faster than that in the cylinders and the flow losses will be correspondingly larger.

Another configuration that is sometimes practically useful is shown in Figure 3.3. In this case, the movement of liquid in the connecting portion is much less than in the cylinders and the natural frequency is determined mainly by the vertical length of the liquid column.

Let the pressure in the large reservoir volume be P. If the left-hand liquid surface is displaced x, the right-hand one will be displaced in the opposite direction, by $(A_1/A_2)x$. The accelerations are related to each other in the same way as the displacements. The pressures across the left- and right-hand columns are $(x_1\rho g - P)$ and $(x_2\rho g - P)$, respectively.

The equations for the acceleration in each tube are

$$\ddot{x}_1\rho h_1 = -(x_1\rho g - P) \tag{3.8}$$
$$\ddot{x}_2\rho h_2 = -(x_2\rho g - P) \tag{3.9}$$

and the displacements and accelerations are related to each other by

$$A_1 x_1 = -A_2 x_2 \tag{3.10}$$
$$A_1 \ddot{x}_1 = -A_2 \ddot{x}_2 \tag{3.11}$$

Substituting (3.10) and (3.11) into (3.9) and subtracting from (3.8) gives

$$\ddot{x}_1 = -x_1 \frac{g(A_1 + A_2)}{(A_1 h_2 + A_2 h_1)} \tag{3.12}$$

and similarly for x_2.

The natural frequency is therefore

$$\omega = \sqrt{\frac{g(A_1 + A_2)}{A_1 h_2 + A_2 h_1}} \quad \text{rad/s} \tag{3.13}$$

or

$$f = \frac{1}{2\pi} \sqrt{\frac{g(A_1 + A_2)}{A_1 h_2 + A_2 h_1}} \quad \text{Hz} \tag{3.14}$$

These reduce to the same equation as (3.2) and (3.3) if the two uprights have equal areas, since the total effective length of the columns is then $h_1 + h_2$.

To summarize the results for displacers, where the restoring force is simply gravity, three cases are given:

1. U-tube configuration, length L_D, constant cross section:

$$\omega = \sqrt{\frac{2g}{L_D}} \quad \text{rad/s}$$

2. U-tube configuration; wide, short cylinders of area A_1; narrower connecting tube of area A_2, length L:

$$\omega = \sqrt{\left(\frac{2g}{L}\right)\left(\frac{A_2}{A_1}\right)} \quad \text{rad/s}$$

3. Reservoir configuration, one upright with area A_1 and length h_1, the other with area A_2, length h_2:

$$\omega = \sqrt{g\frac{(A_1 + A_2)}{(A_1h_2 + A_2h_1)}} \quad \text{rad/s}$$

In all these cases, without going to extremely small or extremely large dimensions, typical operating frequencies will be in the range $\frac{1}{4}$ to 2 Hz (15 to 120 cycles per minute).

TUNING-COLUMN FREQUENCY

The calculation of a natural frequency for the tuning column is slightly more difficult. This is because the forces caused by compression, or expansion, of the gas above the liquid in the machine are not canceled out by acting equally on both ends of the liquid column, as they are in the displacer.

Figure 3.4 shows the configuration of the tuning column. It has a (liquid) length L_t and cross-sectional area A_t. One end is open to the atmosphere. The other end terminates in the working space, which initially has a volume V_m (the volume when the tuning column is at midstroke).

Displacing the water level in the open end of the tuning column

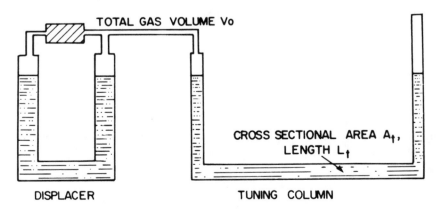

Figure 3.4. Tuning-column configuration with separate cylinders.

downward by an amount x does two things. First, it raises the liquid at the other end by an amount x. Second, as a result of this, it reduces the volume of the gas in the working space by an amount $A_t x$. Both effects give rise to a pressure difference across the tuning column tending to force it back toward the equilibrium position.

If the gas space is isothermal (as we would like it to be in making an ideal Stirling engine) and is initially at a pressure P_m, then the pressure will rise by an amount p where

$$P_m V_m = (P_m + p)(V_m - A_t x)$$

according to the ideal-gas law. Therefore,

$$V_m p = A_t x (P_m + p)$$

Stirling engines usually have a relatively low compression ratio, so that p is generally fairly small compared with P_m. To a reasonable approximation, therefore, $p = A_t P_m x / V_m$.

The pressure difference ΔP acting on the tuning column is therefore

$$\Delta P = \frac{P_m A_t x}{V_m} + 2\rho g x \qquad (3.15)$$

$$\underset{\substack{\text{Pressure} \\ \text{due to gas} \\ \text{compression}}}{} \quad + \quad \underset{\substack{\text{Pressure} \\ \text{due to} \\ \text{extra head}}}{}$$

$$\Delta P = \left(\frac{A_t P_m}{V_m} + 2\rho g \right) x \qquad (3.16)$$

This pressure acts on the liquid in the tuning column, causing it to accelerate.

$$A_t \rho L_t \ddot{x} = -A_t \left(\frac{A_t P_m}{V_m} + 2\rho g \right) \qquad (3.17)$$

therefore,

$$\ddot{x} = -\left(\frac{A_t P_m}{V_m \rho L_t} + \frac{2g}{L_t} \right) x \qquad (3.18)$$

The natural frequency is therefore

$$\omega = \sqrt{\frac{A_t P_m}{V_m \rho L_t} + \frac{2g}{L_t}} \quad \text{rad/s} \qquad (3.19)$$

or

$$f = \frac{1}{2\pi} \sqrt{\frac{A_t P_m}{V_m \rho L_t} + \frac{2g}{L_t}} \quad \text{Hz} \qquad (3.20)$$

The second term is the gravitational one: in the case of the displacer calculations, it was the only term present.

We next look at a configuration (Figure 3.5) that represents a merged-cylinder machine. Actually, the extension from the separate tuning-line calculation we have just done to a merged configuration is very straightforward. The expression for pressure change due to gas compression is the same as before. The liquid level in the displacer, which is assumed to be of larger cross section than the tuning column, does not change as much as the level in the open end of the tuning line. The pressure difference between the liquid surface in the displacer and in the open end of the tuning column is

$$\Delta P = \frac{P_m A_t x}{V_m} + \rho g x + \frac{\rho g x A_t}{2 A_d} \qquad (3.21)$$

Gas com- pression	+ Tuning liquid level lowered	+ displacer liquid level raised

Most of this will act across the tuning column (we can usually ignore the relatively short, wide displacer tubes), and so it becomes a simple matter of using Equation (3.21) in place of Equation (3.16) for the pressure difference along the column. With this substitution,

$$\omega = \sqrt{\frac{A_t P_m}{V_m \rho L_t} + \frac{[1 + A_t/2A_d]g}{L_t}} \quad \text{rad/s} \qquad (3.22)$$

or

$$f = \frac{1}{2\pi} \sqrt{\frac{A_t P_m}{V_m \rho L_t} + \frac{(1 + A_t/2A_d)g}{L_t}} \quad \text{Hz} \qquad (3.23)$$

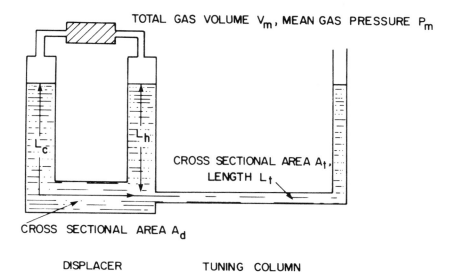

Figure 3.5. Tuning-column configuration with merged cylinders.

If we are striving for greater accuracy and therefore do not wish to ignore the effect of the liquid columns in the displacer cylinder, the length L_t is replaced by $L_t + (A_t/A_d)[L_cL_h/(L_h + L_c)]$ [see, for example, Geisow (1976)].

What if the gas spaces are not all isothermal? The compressibility of a perfectly isothermal gas is equal to V_m/P_m, where V_m and P_m are, respectively, the initial volume and pressure. This compressibility is part of the first term under the square-root sign in Equation (3.23).

The compressibility of a perfectly adiabatic gas is $V_m/\gamma P_m$, where γ is the specific-heat ratio of the gas (Crandall, 1927). For a mixed iso-thermal-adiabatic volume, the situation is much more complicated and the relation between pressure and volume depends on whether the gas is leaving the adiabatic space for the isothermal space, or vice versa (Rios, Smith, and Qvale, 1969). However, this has a fairly small effect on the average compressibility, so we simply calculate the overall com-pressibility of the working gas from the weighted average of the mean isothermal and adiabatic volumes, V_i and V_a.

$$\frac{\Delta P}{\Delta V} = \frac{-P_m}{V_i + V_a/\gamma} \tag{3.24}$$

Substituting this into Equation (3.23) gives an approximate formula for the natural frequency of the tuning line in a machine with mixed isothermal and adiabatic spaces.

$$f_t = \frac{1}{2\pi} \sqrt{\frac{1}{L_t}\left[\frac{\pi R_t^2 P_m}{\rho(V_i + V_a/\gamma)} + g\left(1 + \frac{R_t^2}{2R_d^2}\right)\right]} \qquad (3.25)$$

R_d is the diameter of the displacer tube.

In many cases, the last two terms are rather small and can be neglected. In any case, Equation (3.25) can be rearranged to give a simple relation between L_t and R_t for any given frequency of operation, f.

$$L_t = \frac{\left\{\dfrac{\pi R_t^2 P_o}{\rho[V_i + V_a/\gamma]}\right\} + g\left[1 + \dfrac{R_t^2}{2R_d^2}\right]}{4\pi^2 f^2} \qquad (3.26)$$

POWER OUTPUT

We could calculate the power output using the Schmidt formula, but for our present circumstances it will be better to use a simplified expression (Cooke-Yarborough, 1974), because it separates the effects of various parameters more clearly. One formula based on Cooke-Yarborough's approximation (West, 1971) for a machine with separate displacer and tuning column cylinders gives

$$W_o = P_m V_o f \pi \frac{V_e}{4V_m} \frac{T_e - T_c}{T_e + T_c} \sin\theta \qquad (3.27)$$

where V_o = volume swept out by the surface of the tuning column
V_e = volume swept out by either surface of the displacer
V_m = midstroke volume
P_m = mean pressure
T_e and T_c = temperatures of the hot and cold spaces, respectively
θ = phase angle between the displacer and tuning column
f = frequency of operation

Ideally, we try to get the two movements fairly near to 90° out of phase, i.e., $\theta = 90°$, in order to maximize the value of sin θ. Because sin θ does not change very rapidly with θ in this range, assuming that sin $\theta = 1.00$ will give rise to an error of less than 15 percent if the phase angle is 120° instead of 90°. Martini (1978) has collected data on solid piston Stirling engines and observes that actual net power output (i.e., after losses) is usually 0.3 to 0.4 times the output calculated ideally. The very limited experience available with well-designed liquid piston Stirling engines is consistent with this. We therefore use the figure 0.3 as a suitable correction factor, recognizing that at lower temperatures or for small machines (where various losses become more important—see Chapter 5) the formula will tend to overestimate the power available.

$$W_{\text{net}} \simeq 0.3 P_m V_o f \frac{\pi}{4} \frac{V_e}{V_m} \frac{T_e - T_c}{T_e + T_c}$$

$$\simeq 0.25 P_m V_o f \frac{V_e}{V_m} \frac{T_e - T_c}{T_e + T_c} \tag{3.28}$$

Walker and Agbi (1973a, b) showed, by means of an idealized theoretical model, that the use of a two-phase, two-component working fluid can increase the specific power output by a factor of 2 or 3. Such an effect can easily be realized in a liquid piston engine, simply by allowing some of the liquid in the hot cylinder to evaporate. Indeed, unless specific precautions to avoid evaporation are taken—such as placing an insulated float on the liquid surface (West and Geisow, 1975) or using a low-vapor-pressure liquid—the effect will take place naturally but in a somewhat uncontrolled way.

This can be an important advantage for the Fluidyne, especially in small machines where the losses are relatively higher (because of the increased flow resistance in narrow tubes and the increased surface-to-volume ratio) and the greater output resulting from evaporation may be essential to the operation of the engine.

Of course, the increased power output is obtained at the expense of an increased heat input to provide the latent heat of evaporation, although other heat losses may be large enough in very small machines—which, for demonstration purposes, are often constructed without thermal insulation—that the latent heat is not significant. The effects of evaporation are discussed in greater detail in Chapter 8.

In this chapter, the major design factors of the Fluidyne engine have been discussed in an idealized, but not too unrealistic, manner. In the next chapter, the actual construction of some small, working machines is described. Later, some effects that must be taken into account in making more advanced power and loss calculations are discussed.

Chapter 4
Practical Examples

One of the advantages of the liquid piston system is constructional simplicity. A liquid piston alway fits its cylinder eactly; no piston rings are needed for sealing; no bearings are required to support rotating or oscilating components. Consequently, Fluidyne engines can be quickly and easily built with the simplest of tools and materials.

PLASTIC ENGINE

One of the most straightforward engines to construct is the plastic machine shown in Figure 4.1. It is a liquid feedback mechanism made by gluing $\frac{3}{8}$-in-thick clear acrylic plastic strips between $\frac{1}{8}$-in-thick cover plates. Heat input is provided by a small 20-Ω 3-W wire-wound resistor in the hot cylinder.

Evaporation is very important in this engine: if the water in the hot cylinder is replaced by oil, the engine will not run. The length of the tuning line is chosen for constructional simplicity; it is, in fact, very much shorter than the ideal length calculated according to Equation (3.26). Nevertheless, the machine will run quite well with 9 V on the heater, corresponding to a power of 4 W. Better results are obtained if sugar is added to the water (30 percent by weight) and the input voltage is raised to 12.6 V, or a power of 8 W. The sugar increases the viscosity of the water and results in a more favorable flow pattern (see Chapter 6 for an explanation of this effect). Note that in either case, the input power exceeds the rating of the resistor which is, however, able to survive if the water level is not allowed to fall too low, so that the resistor is in effect water cooled.

Naturally it is essential, for safety reasons, to use only a low voltage on the heater of the machine or, indeed, on any component with which an experimenter or onlooker might come into contact.

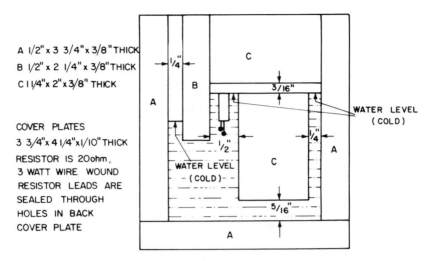

Figure 4.1. Plastic machine made by glueing clear Acrylite acrylic strips between cover plates (drawn to scale).

The cylinder walls are also cooled by the water, and that is why plastic, a relatively low melting-point material, can be used in the construction.

WOODEN ENGINE

For some people, wood may be a more convenient material because it is somewhat easier to cut and drill than is plastic. The wooden engine shown in Figure 4.2 is larger than the plastic machine, with an output stroke of approximately 10 cm, or 5 cm^3. The original version was made from $\frac{3}{4}$-in-thick redwood, a very pleasant material to work with, although there seems no reason why exterior grade plywood could not be used. Assembly is with brass screws and a good-quality waterproof glue (which makes the screws perhaps unnecessary).

The heater element is a 10-Ω 10-W resistor (available in the United States from Radio Shack Stores, catalog number 271-132). It is fed from a 12.6-V transformer, so that the input power is approximately 15 W. The electrical connections are small bolts passing through clearance holes in the wood.

Thin aluminum foil is folded, concertina-like, into four or five corrugations and placed in the cold cylinder. This isothermalizer provides enough surface area for the incoming gas to give up its heat. On the upstroke in the cold cylinder, the aluminum strips are cooled by the rising water (West and Geisow, 1975).

The wooden cylinder walls are reasonably effective thermal insulators (otherwise the hot end of the machine could not be heated with an input of only 15 W). Accordingly, some way must be found for providing extra cooling of the water at the cold end of the machine. One way of doing this is shown in Figure 4.2: as the pressure inside the engine rises, some warm water is forced from the cold cylinder into the cooling water reservoir, mixing with the cooler water already there, by compressing the air in the reservoir. On the downstroke of pressure, water from the reservoir is drawn back into the cylinder. The can or jar forming the reservoir has a large enough surface area and thin enough walls to provide adequate cooling by convective heat transfer to the atmosphere.

The tuning line is made from $\frac{5}{16}$-in clear plastic tubing, topped by a

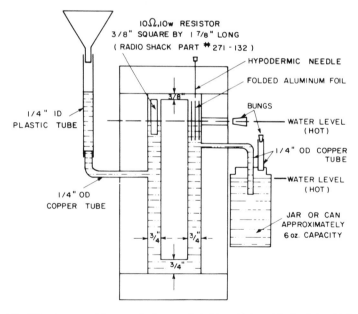

Figure 4.2. Wooden machine made by sandwiching $1\frac{1}{2}$ in wide by $\frac{3}{4}$ in thick wooden strips between 13 in high by 6 in wide by $\frac{3}{4}$ in thick wooden cover plates (drawn to scale).

small funnel. The funnel makes it easy to fill the engine, and also prevents an overflow should the pressure in the machine increase too much. The hypodermic needle, sealed in place with silicone rubber, is a refinement and not essential to the operation of the engine: its purpose is to provide a slow leak that keeps the average pressure inside the machine close to atmospheric and thus makes the average level of the water in the cylinders and tuning line equal. Without such a leak, it will be necessary to remove the level plug from time to time in order to equalize the pressures.

Considerable space has been devoted to the description of this engine and the plastic one because both are easier to make than any other configurations known at present. They thus represent a good starting point for anyone wishing to become involved in liquid piston engine work. Moreover, there are still worthwhile and original experiments to be done with them: determining how the amplitude of the oscillation in the tuning column of the wooden machine varies as its length is increased; finding out what happens if the tuning line is inserted higher up or lower down in the side of the hot cylinder; measuring what happens if the viscosity of the water is increased by adding sugar; seeing if there is enough power available to drive a small pump; ascertaining if this form of construction can be used to make a much larger machine. These questions, and others, could be answered experimentally through undergraduate or even high school projects.

GLASS DEMONSTRATION PUMP

Most of the early working Fluidynes were made from glass. For those with access to glassblowing facilities, glass offers many advantages. Borosilicate glass (e.g., Pyrex) can withstand high temperatures and high-temperature gradients. It is not permeable to water or air, and it is transparent. For many people, including this author, who have had the opportunity to work with a skilled and creative glassblower, glass is the material of first choice.

The approximate dimensions of one of Harwell's small demonstration models of a pump and engine system, devised by David Herbert, are given in Figure 4.3. This machine has been exhibited many times and has proved to be reliable and eye-catching. Heating is normally provided by a 16-W quartz halogen projector bulb focused onto the hot

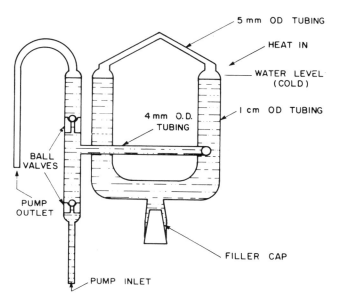

Figure 4.3. Small glass Fluidyne pump devised by David Herbert (drawn to scale).

cylinder. Also demonstrated was that models of this type, and larger glass machines, can easily be solar-powered using an inexpensive plastic Fresnel lens. Best results are usually obtained if the sunlight is focused onto the hot cylinder at approximately the mean water level.

With either a lamp or sunlight, the amount of heat absorbed at the hot end of the machine can be increased by placing a reflector at the back of the cylinder, or by blackening one of the surfaces. Adding, say, ink to the water in order to render it opaque would also increase the absorption.

These engines can also be heated with a small alcohol burner, or even a candle. A short section of aluminum or steel channel is used to direct the heat of the flame onto the cylinder (Figure 4.4). This arrangement, proposed by John Geisow, provides a simple, inexpensive, portable heat source. The borosilicate glass can easily withstand the large temperature gradient between the water-cooled interior wall of the cylinder and the flame-heated exterior. A spirit lamp, i.e., an alcohol burner, is preferred to a candle, because the latter gives rise to much soot and soon spoils the appearance, at least, of the engine.

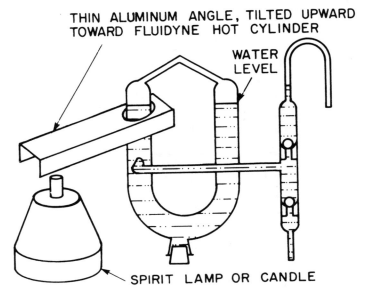

THIN ALUMINUM ANGLE, TILTED UPWARD
TOWARD FLUIDYNE HOT CYLINDER

WATER
LEVEL

SPIRIT LAMP OR CANDLE

Figure 4.4. Flame-heating for small glass Fluidyne pump.

LABORATORY SYSTEM

One of the earliest designs, and one which has provided a starting point
for several other experimentalists seeking to build their own machines,
was a combined engine and pump (West, 1971). The details are given
in Figure 4.5. This engine was constructed from copper and brass pipe
fittings, and the pump used 1-in valves consisting of rubber-faced metal
discs lightly spring-loaded onto a circular opening (aircraft fuel-line
valves). An input power to the electric heater of 250 W gave a maximum pumping rate of 370 L/h (100 U.S. gal/h), for a small pumping
head. By raising the input power to 530 W, 100 U.S. gal/h could be
pumped against a head of 1.6 m (5ft 3in). The maximum efficiency was
about 0.35 percent, and it is now known that the effects of evaporation
from the water in the hot cylinder contributed substantially both to the
high power output and the low efficiency.

FRUIT-JAR MACHINE

A smaller design for combined engine and pumping systems, also published by Harwell, is based upon a fruit jar (Figure 4.6). The pump uses

ball valves that can conveniently be made from ball bearings or glass beads. The glass U tube forming the cylinders and regenerator may be difficult to acquire without access to a glass-blowing shop. In its place a separate tube could perhaps be used for each cylinder and the cylinders connected to each other by a length of plastic tubing. This machine will pump 5 U.S. gal/h (Mosby, 1978).

The fruit jar machine operates easily and without critical adjustments as a free running engine, i.e., with the pump disconnected. However, in some cases the water level in the cylinders and the length of tuning line inserted into the hot cylinder may require careful adjustment for successful operation with the pump series-coupled as shown in Figure 4.6. Some simple, but very useful, experimental work could be done by comparing the behavior with the pump connected in the parallel and series modes described in Chapter 2.

One way of providing heat to these engines, which has also been used successfully with several other machines, is by means of a hot-air blower. A hair dryer generally does not seem to give a high enough

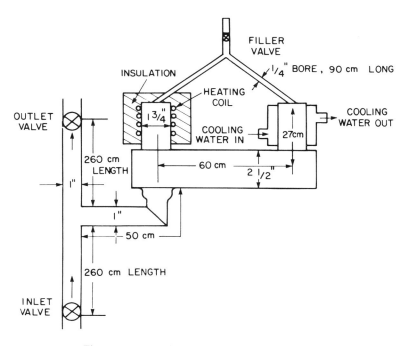

Figure 4.5. Fluidyne pump. (After West, 1971.)

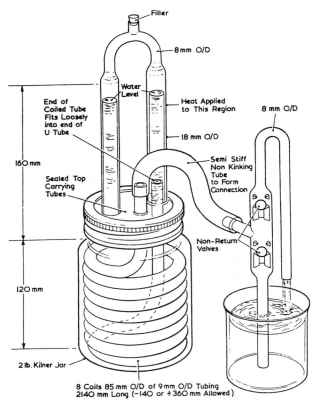

Figure 4.6. Fruit-jar machine for home construction. (Copied from AERE Harwell brochure.)

temperature, but the blowers (similar in appearance) that are used in the laboratory and on the production line provide a hotter air stream.

ISOTHERMALIZERS

As indicated earlier, in all but the smallest machines the cold cylinder should be subdivided into gaps of no more than several millimeters in order to get good heat transfer. This would be necessary in the hot end of the machine also if no evaporation were permitted, unless heat is supplied to the gas before it enters the cylinder.

One way of achieving this subdivision in the cold cylinder has already been mentioned: aluminum sheet is folded into a concertina-like shape and fixed inside the cylinder. In a wooden machine, it can be nailed in place. Another simple solution, based on a suggestion by Walker, would be to use drinking straws packed into the cylinder: this is an inexpensive way of buying a lot of holes, which is the basic objective. A honeycomb material, such as Aeroweb, can be used, which has the advantage of having all the holes identical; such is not the case for a collection of round tubes, because of the spaces between them. In any case the tubes or slots should be wide enough (several millimeters) so that surface tension does not hold the water, or other displacer fluid, within them.

The effect of this subdivision is to make the gas in the cylinder behave more or less isothermally. By contrast, gas that is in a large cylinder, or driven by a very rapidly oscillating piston, will behave adiabatically. Martini, Hauser, and Martini (1977) have stressed the advantages of isothermal cylinders and have proposed several ways of achieving this state even in high-speed solid piston engines. Following Martini's practice, we shall refer to the devices for achieving isothermal or near isothermal gas behavior in the cylinder as isothermalizers.

REGENERATORS

Much attention is paid, in conventional Stirling engines, to achieving good regeneration. This is important in a Fluidyne also, if high efficiency is desired. (The reader is reminded that the regenerator does not directly affect the output power but it does directly increase the efficiency.) What experimental evidence is available suggests that in liquid piston machines with substantial evaporation from the hot cylinder, regeneration is not very effective. In such machines, a single connecting tube between the hot and cold cylinders is often used.

If better regeneration is desired—for example, when evaporation is suppressed so that the engine operates more nearly as a true Stirling machine—a more sophisticated arrangement may be necessary.

At low temperatures, drinking straws can be used as a regenerator; and at higher temperatures, metal tubes or honeycomb will serve. However, annular regeneration is theoretically the most efficient configuration for laminar flow, because it has the highest ratio of heat-transfer effect to flow loss; that is, it provides the best compromise between thermal efficiency and power loss due to viscous effects. In many cases, especially for the larger machines with a cylinder diameter of several inches, an annular regenerator, with a gap of a few millimeters, fits well into the overall dimensions of the engine. This shape is also less prone than is a tube to blockage by surface tension. However, it is very important that the annular gap be uniform around the circumference of the cylinder, otherwise the regenerator effectiveness will be greatly reduced.

As in conventional Stirling machines, regenerators can also be made from wire screens, lengths of wire, or beads. However, as long as there is a chance of evaporation or splashing, small gaps or interstices should be avoided lest surface tension holds drops of condensate in place and causes partial regenerator blockage.

Recent developments in silicone adhesives and sealing compounds have simplified the Fluidyne constructor's task considerably. Some of these compounds—for example, General Electric's Hi-Temp Instant Gasket, retailed inexpensively as an automotive product—will withstand up to 590 K (600°F), which is a not unreasonable temperature for the hot end of a Fluidyne.

SELF-STARTING

Liquid feedback Fluidyne engines are usually self-starting: when the temperature is raised beyond some threshold level, the liquid begins to oscillate of its own accord. Rocking feedback machines usually require an initial push to overcome static friction at the pivot: presumably a frictionless suspension—flexure or spring—could be self-starting. When a pump is attached to the engine, some starting assistance, such as manually rocking the engine, may be needed. In the case of a series pumping arrangement (Figure 2.16), the reason is easy to see: until the oscillation builds up enough pressure to open the valves, no liquid movement can take place in the tuning line; but until there is liquid movement in the tuning line, there is no feedback and hence no oscillation. In practice, this is not usually a problem, at least for small machines, because the lack of oscillation allows the temperature to rise until some local boiling takes place in the hot cylinder. The resulting pressure rise opens the valve and oscillation begins. This process takes heat from the cylinder and reduces the temperature below the boiling point, but by this time the cycle is established and continues. Parallel and gas-coupled pumping systems are usually self-starting even without evaporation and boiling.

The best way to learn something about Fluidyne engines is to build one and operate it. This chapter indicates that this can be done fairly simply. In this respect, one further point should be made: in the author's experience, it is quite difficult to construct a small Fluidyne that cannot be made to work. Neither the dimensions nor the constructional details are very critical. To make a Fluidyne that works efficiently and gives a high output is a different matter, as we shall see.

Chapter 5
More Advanced Power Calculations

The liquid feedback Fluidyne engine has no rotating or sliding solid parts, and therefore no mechanical friction. It does, however, suffer from the viscous and other losses associated with flowing fluids, especially flowing liquids. In common with other Stirling engines, it also suffers from the fact that the cylinders and other gas spaces are, in general, neither perfectly isothermal nor perfectly adiabatic.

VISCOUS FLOW LOSSES IN OSCILLATING FLUIDS

Most of the fluid-flow problems met by engineers concern steady flow in long ducts or channels. This is not the case for the Fluidyne designer, since he is dealing with an oscillating system, which modifies the flow behavior profoundly. Elrod was the first to point out that flow patterns in the Fluidyne would be quite different from unidirectional, well-developed flow and may instead be similar to the behavior of a fluid close to the beginning of a pipe section, known as "entry flow."

The problem of oscillating flow was studied in the last century (Rayleigh, 1896) by Kirchhoff and other physicists interested in the behavior of sound waves in tubes or porous materials. In 1975, Ryden analyzed oscillating flow in the Fluidyne system, but his work has not been published; we shall rely instead on the formulation given by Crandall (1927).

To determine viscous losses due to oscillating flow in a tube, Crandall shows that we must first decide whether the tube is to be treated as narrow or wide. It is "wide" if the dimensionless quantity $R_t \sqrt{2\pi f \rho / \eta}$, which we call R^*, the radius parameter, is much greater than unity (R_t is the tube diameter, f the frequency of the oscillation, and ρ, η are the density and viscosity, respectively, of the liquid). Narrow tubes, even with oscillating flow, obey the usual Poiseuille flow laws.

Substituting the parameters for water at room temperature, and recalling that the frequency of operation of a Fluidyne will usually be around 0.65 Hz, we find that the radius parameter becomes, in SI units, $R_t\sqrt{2\pi \times 0.65 \times 10^3/0.001} \simeq 2000R_t$. Thus all cylinders and isothermalizers with a radius of a few millimeters (10^{-3} m) or more can be treated as wide when one calculates viscous effects of water flow.

The resistance coefficient is defined as the pressure drop per unit length divided by the mean flow velocity. In normal laminar flow it is called the Poiseuille resistance coefficient and is equal to $8\eta/R_t^2$. Crandall shows that for nonturbulent oscillating flow in wide tubes, the resistance coefficient is $\sqrt{2\rho\omega\eta}/R_t$.

$$R = \frac{\sqrt{2\rho\omega\eta}}{R_t} \tag{5.1}$$

Under these circumstances the velocity profile across the tube is not, as in Poiseuille flow, parabolic. Rather, at any instant the velocity is constant across most of the tube diameter, decreasing to zero across a relatively thin boundary layer close to the tube wall (Figure 5.1). The liquid flows as if it were almost solid in the central region or core of the tube. Note that in this regime the pressure drop for a given flow rate increases only as the square root of the viscosity and does not increase as rapidly with decreasing tube size as it does in Poiseuille flow.

Laminar flow with $R^* \leq 4.5$ is described well by the Poiseuille formula and for $R^* \geq 8.5$, the solid core-type behavior is seen. Intermediate values of R^* can lead to a much more complicated flow pattern (Chan and Baird, 1974).

In any kind of flow, the instantaneous rate of loss of power is equal to the volume flow rate multiplied by the pressure drop. This is easily calculated from the resistance coefficient which, by definition, is the pressure drop per unit length divided by the mean flow velocity. The volume flow rate is simply equal to the mean flow velocity multiplied by the cross-sectional area of the tube. Therefore

$$\Delta P \dot{V} = \frac{RL_t\dot{V}}{\pi R_t^2} \, \dot{V} \tag{5.2}$$

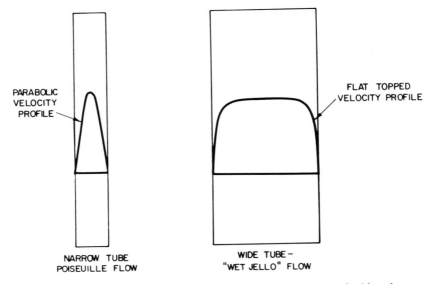

Figure 5.1. Velocity profiles for oscillating flow in narrow and wide tubes.

For the Fluidyne machine, we should use Crandall's expression for the resistance coefficient. Substituting this in Equation (5.2) yields

$$\Delta P \dot{V} = \sqrt{2\rho\omega\eta} \, \frac{L_t \dot{V}^2}{\pi R_t^3} \tag{5.3}$$

Equation (5.3) shows that the instantaneous value of the rate of power loss is proportional to the square of the volume flow rate. The average value over a whole cycle is therefore dependent on the mean square of the volume flow rate.

The mean square value of any sinusoidal function is equal to half the peak square value, and the peak value of $\dot{V}$ is simply equal to the volume amplitude (or half the volume stroke) multiplied by the angular frequency. Therefore

$$\overline{\Delta P \dot{V}} = E_{\text{viscous}} = \frac{\pi \sqrt{\pi \rho f \eta} L_t f^2 V_o^2}{R_t^3} \tag{5.4}$$

Mean   Viscous
value   flow loss

In using Equation (5.1) and its derivatives, we have implicitly assumed that the flow is not turbulent. The available data on the critical Reynolds number for the development of turbulence in oscillating flow clearly show that the transition may take place at much higher Reynolds numbers than is the case for steady flow. Park and Baird (1970) derive a formula for the critical Reynolds number in oscillating flow; they also note, from their experimental results, that when surface and end effects are negligible, their equation underestimates the critical Reynolds number by about 50 percent. With this adjustment, and a change of symbols, their equation reads

$$R_e(\text{critical}) \simeq 375\left(\frac{R_i^2 \omega \rho}{\eta} \right)^{2/3} \tag{5.5}$$

Table 5.1 shows the important parameters relating to oscillating flow effects in different regions of a typical Fluidyne. For comparison, the same parameters are calculated for a conventional Stirling engine, the GPU-3.

According to Table 5.1, it appears that the laminar flow approximation can be valid, or nearly valid, for a Fluidyne of reasonable dimensions. However other more recent results on oscillating liquid flow (Chan and Baird, 1974) seem to indicate that Equation (5.4) may underestimate the laminar viscous flow losses by a factor of 3 or more. Furthermore, some of the Chan and Baird results appear to imply that the flow may be fully turbulent under conditions where calculations based on the Park and Baird equation would predict laminar flow patterns. The Chan and Baird results taken at high amplitudes and low frequencies—up to 0.4 m at 0.25 Hz and 0.45 Hz in a 50-mm-diameter steel tube—differ by as much as a factor of 5 from those calculated on a laminar flow basis, and the discrepancy was wider at the higher frequency.

The foregoing, somewhat extended, discussion has revealed that the effects of oscillating flow in the Fluidyne liquid columns are very far from being completely understood. Analyzed with our present theories, the published results appear not to be fully consistent with each other. However, it is clear that in the absence of turbulence the solid-core approximation is a much better representation of Fluidyne liquid-flow

Table 5.1. Flow Characteristics in Various Engine Components.

	FLUIDYNE			GPU-3	
	OUTPUT COLUMN	DISPLACER	ISOTHERMALIZER*	HEATER	COOLER
Fluid	Water	Water	Air	Hydrogen	Hydrogen
Pressure, MPa	0.1	0.1	0.1	6.5	6.5
Swept Volume, 10^{-6} m^3	2000	1400	2.2†	2.9†	0.4†
Frequency, Hz	0.6	0.6	0.6	50	50
Temperature, K	293	293	300	950	350
Density, Kg/m^3	1000	1000	1.2	1.6	5.4
Viscosity, 10^{-5} ns/m^2	100	100	1.9	2.3	0.9
Radius, mm	25 31	75	3	1.5	0.5
Radius parameter	50 60	150	1.5	7.0	6.9
Reynolds number	95,000 77,000	23,000	60	13,000	45,000
Critical Reynolds number	67,000 88,000	290,000	2,000‡	5,000	4,900
Flow regime	Core flow turbulent	Core flow laminar	Poiseuille laminar	Intermediate turbulent	Intermediate turbulent

*Six-millimeter-diameter tubes placed axially in the cold cylinder.
†Swept volume per tube.
‡Critical value for Poiseuille flow.

behavior than is the assumption of Poiseuille flow, at least for tubes wider than a few millimeters. At sufficiently high Reynolds numbers, it may be that the well-known turbulent-flow correlations between velocity, fluid properties, tube size, and pressure drop also apply to the oscillating flow case and can be used unchanged; but this is by no means certain, nor are the limits of the various flow regimes very well-defined. There is a great need for some measurements under conditions of tube size, frequency, and amplitude that are typical of real Fluidynes.

KINETIC FLOW LOSSES

In addition to the viscous losses, there are the so-called minor pipe losses that occur when the fluid must change speed or direction, as at a bend or pipe exit. The applicability to oscillating flow of the standard formulas for minor pipe losses, such as those given by the Crane Company (1957), does not seem to have been established. There is certainly a need for experimental data on this question too. In any case, we have no option at present but to assume that the standard formulas are at least approximately valid.

Generally speaking, the pressure drop (ΔP) caused by a minor pipe loss can be expressed as a factor, which depends on the type of obstruction, multiplied by the velocity head:

$$\Delta P = K \tfrac{1}{2}\rho v^2$$

The total pressure drop due to all the bends, constrictions, enlargements, etc., in the system is obtained by summing the contributions from each of them individually; this assumes that the obstructions are far enough apart that the flow disturbance for each is not modified by the others. Table 5.2 quotes typical values of K for some representative flow obstacles.

The flow velocity is equal to the volume flow rate $\dot{V}$ divided by the cross-sectional area of the tube. The instantaneous rate of loss is $\Delta P \dot{V}$.

$$\Delta P \dot{V} = \Sigma K \frac{1}{2} \rho \left(\frac{\dot{V}}{\pi R_t^2} \right)^2 \dot{V}$$
$$= \frac{\Sigma K \rho \dot{V}^3}{2\pi^2 R_t^4}$$

(5.6)

Table 5.2. Minor Pipe Loss
Coefficients.

ELEMENT	K
90° smooth bend	0.15–0.25
90° mitre bend	1.0
Sharp edged contraction	0.5
Sharp edged enlargement	1.0
15° conical enlargement or contraction	0.2

This is evaluated in the same way as Equation (5.3), except that the mean cube value of a sinusoidal function, taking no account of sign, is 0.42 times the peak value of the cube:

$$\text{Minor pipe loss} = E_k \simeq 0.42 \; \Sigma K \; \frac{\pi \rho f^3 V_o^3}{2R_t^4} \qquad (5.7)$$

Note that the minor pipe losses increase with the cube of the frequency and the cube of the swept volume; for this reason, in Fluidyne design they are usually called kinetic losses or cube-law losses. Note also that as tube diameter is decreased, the kinetic losses increase more rapidly than do the viscous losses.

GAS-FLOW LOSSES

Some further points can be made about the gas-flow losses in the regenerator and in the heat exchangers or isothermalizers. In this case, assuming air at 400 K and atmospheric pressure,

$$R_t \sqrt{2\pi f \rho/\eta} \simeq R_t \sqrt{(2\pi \times 0.65 \times 0.88)/2.3 \times 10^{-5}}$$
$$\simeq 400 \; R_t$$

For tubes with a diameter of a few millimeters, such as we might use for heat exchangers or regenerators, this quantity is about 1 (see Table 5.1). Therefore, if the flow is laminar, it can be approximated by normal Poiseuille flow, for which the resistance coefficient is $8\eta/R_t^2$.

Even with this simplification, the proper calculation of gas-flow losses is complex, especially in the regenerator, because the temperature and mass flow rate of the gas vary substantially down the length of the regenerator. It is, however, a standard problem common to all Stirling machines, and a description of it is given in Martini's *Stirling Engine Design Manual* (1978). We give only an approximate method, suitable for a rough estimate of these losses.

In this approximation, we assume that the volume flow rate in the regenerator (or heat exchangers) is sinusoidal and equal in amplitude to the rate of change of volume in either cylinder. This is approximately $V_o/\sqrt{2}$ for phase angles close to $90°$. For the regenerator, we use the properties of the gas at the mean pressure of the engine and at a temperature halfway between the hot and cold temperatures. The regenerator is assumed to be formed from N ducts, each with a diameter equal to or equivalent to R_r and of length L_r. The total volume flow rate through each duct, $\dot{V}$, is therefore $\dot{V}_o/(\sqrt{2}N.)$

Once again, the instantaneous rate of loss of energy is $\Delta P \dot{V}$, and from the definition of the resistance coefficient R, we have

$$\frac{\Delta P/L_r}{\dot{V}/\pi R_r^2} = R$$

therefore

$$\frac{\Delta P/L_r}{\dot{V}_o/\sqrt{2}\pi R_r^2 N} = \frac{8\eta}{R_r^2}$$

therefore

$$\Delta P = \frac{8\eta L_r \dot{V}_o}{\pi\sqrt{2}R_r^4}$$

therefore

$$\Delta P\dot{V} = \frac{4\eta L_r \dot{V}_o^2}{\pi N^2 R_r^4} \tag{5.8}$$

This is the instantaneous rate of loss of energy in each of the N ducts. The average total loss is half the peak value in each duct multiplied by the number of ducts.

$$E_{\text{viscous}} = \frac{N}{2}\frac{4\eta L_r}{\pi N^2 R_r^4}\left(\frac{2\pi f V_o}{2}\right)^2$$

$$= \frac{2\pi\eta L_r f^2 V_o^2}{N R_r^4} \tag{5.9}$$

This gives a very approximate method for calculating the viscous losses in the regenerator ducts. What about the kinetic losses? The pressure drop due to the minor pipe loss is proportional to the gas density and to the square of the gas velocity in the regenerator. For a given volume flow rate (i.e., for a given piston stroke), the power loss is proportional to the pressure drop.

The ratio of the gas velocity in the regenerator entrance and the liquid velocity in the displacer is approximately equal to the ratio of regenerator area and cylinder area. The kinetic losses in the regenerator will therefore be much less than in the displacer liquid if

$$\frac{(\text{Gas density})/(\text{regenerator area}^2)}{(\text{Liquid density})/(\text{displacer area}^2)} \ll 1$$

Taking the density of air in the regenerator as approximately 1 kg/m³ and of the displacer liquid as 1000 kg/m³, this inequality is fulfilled if

$$\frac{R_d^4}{R_r^4} \ll 1000 \tag{5.10}$$

or
$$R_r \gg 0.18 R_d \tag{5.11}$$

As long as the cross-sectional area of the regenerator is more than about 5 percent of the cylinder area, then a design giving acceptable kinetic losses in the displacer will give acceptable kinetic losses in the regenerator also.

Finally, we mention briefly a source of loss that may be expected to occur when a fluid in the core-flow regime enters a region—such as the cold-cylinder isothermalizers—that is subdivided into smaller ducts with very thin walls. In this case the mean flow velocity is unchanged. However, the proportion of the fluid that is in the relatively slowly moving boundary layer increases, and indeed the radius parameter R^* may fall to the point at which the flow approaches the Poiseuille velocity profile. In any case, for a given mean velocity, the mean square velocity and hence the kinetic energy of the flow will be different in the narrow ducts from the preceding undivided tube. This change in kinetic energy may give rise to a pressure drop and therefore to a cube-law flow loss. Also, the viscous loss per unit length will be higher in the narrower ducts.

EFFECT OF ADIABATIC CYLINDERS

The Schmidt analysis of Stirling machines assumes, *inter alia,* that the gas in the cylinders behaves isothermally. For modern, high-speed Stirling engines, this is usually not true; and, indeed, the gas behavior in the cylinders is more nearly adiabatic than isothermal. The heat-exchange spaces, on the other hand, are almost isothermal. The combination of isothermal and adiabatic volumes leads to a loss mechanism that has the effect of reducing the power output and the thermodynamic efficiency. The effect is well known in practice and from computer simulations. Recently the classical analysis of the cycle has been extended to include, in closed form solutions, some of the effects due to an adiabatic cylinder (West, 1980 and 1982a).

A computer code written for a small personal computer is available to calculate the power output, input, and efficiency of an ideal machine in which either or both cylinders behave adiabatically (West, 1979). In the case of the Fluidyne, we have already seen that the cold cylinder can usually be isothermalized, and so the following discussion centers around machines in which only the hot cylinder is adiabatic. As an example, we take an alpha configuration machine (Figure 5.2) with the base parameters shown in Table 5.3.

To make a comparison between the behavior of a machine with both cylinders isothermal and one with the hot cylinder adiabatic, we must first decide what is to be held invariant. The mean pressure? The maximum pressure? The mass of working fluid? For the Fluidyne pump, it is convenient to hold the mean cycle pressure constant because this is easy to accomplish in practice. All calculations are therefore carried out at a mean pressure of 0.1 MPa, or approximately 1 atm.

The results are shown in Figures 5.3 and 5.4. The first thing to note is that the output power from an all-isothermal engine does not fall to zero until there is no temperature difference between the two cylinders. However, use of an adiabatic hot cylinder leads to an internal power loss, even in an engine that is otherwise ideal, and no power is available until, in this particular case, the heater temperature is raised to 60°C. There is an actual power loss; this is shown by the fact that not only the power output but also the efficiency is reduced. Physically, the mechanism is that as the pistons move, gas from the heater is mixed with gas from the cylinder, which is at a different temperature because it is behaving adiabatically. Mixing fluids at different temperatures is an

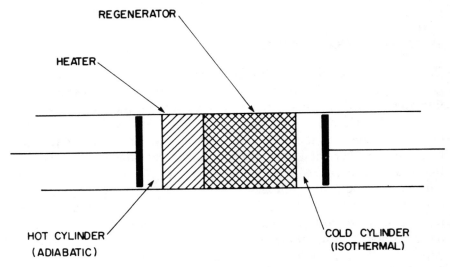

Figure 5.2. Alpha-configuration engine with adiabatic hot cylinder and isothermal cold cylinder.

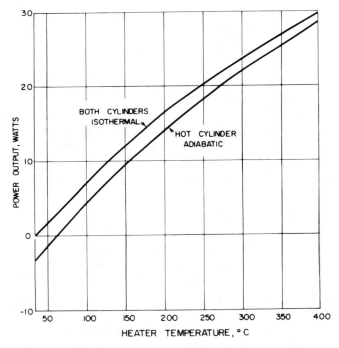

Figure 5.3. Power output as a function of heater temperature.

Table 5.3. Basic Engine Parameters for Ideal Output
Calculation.

PARAMETER	VALUE
Hot-cylinder swept volume	10^{-3} m^3 (1 L)
Hot-cylinder unswept volume	10^{-4} m^3 (100 cm^3)
Cold-cylinder swept volume	10^{-3} m^3 (1 L)
Cold-cylinder unswept volume	10^{-4} m^3 (100 cm^3)
Heater volume	3×10^{-4} m^3 (300 cm^3)
Regenerator volume	7×10^{-4} m^3 (700 cm^3)
Cooler volume	0
Compression space temperature	308 K (35°C)
Frequency	1 Hz
Phase angle	90°
Working fluid	Air

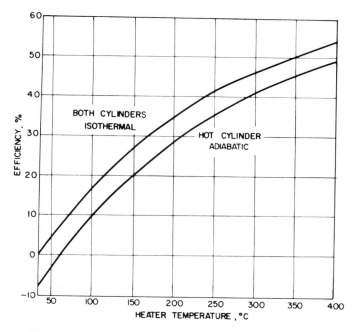

Figure 5.4. Efficiency as a function of heater temperature.

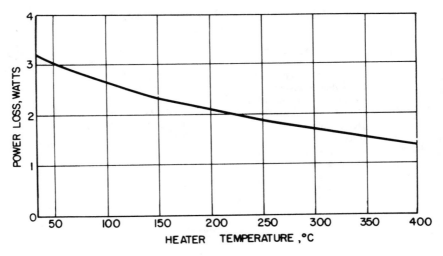

Figure 5.5. Power loss due to adiabatic hot cylinder as a function of heater temperature.

irreversible process, in thermodynamic terms, so that not all the energy involved can be recovered.

At high operating temperatures, the effect is relatively small, but it is absolutely and proportionately larger at lower heater temperatures (Figure 5.5).

CORRECTED SCHMIDT ANALYSIS

Once the temperature difference, ΔT_{min}, required to give a zero output power for the adiabatic cylinder machine is known [25 K in this example (see Figure 5.3)], rather accurate estimates of the power output at other temperatures can be made using the Schmidt formula with an effective heater temperature reduced by ΔT_{min} (see Table 5.4).

TRANSIENT HEAT-TRANSFER LOSS

As the gas in the machine is expanded and compressed, its temperature tends to fall and rise. However, the gas immediately adjacent to any solid surface (such as the cylinder walls, regenerator matrix, etc.)

hardly changes at all if the thermal capacity and conductivity of the wall are much greater than that of the gas. Consequently, the gas near the wall will tend to behave isothermally, whereas any gas far removed from the wall tends to behave adiabatically. There is therefore a continuously varying temperature gradient between the bulk of the gas and the adjacent walls; heat conduction down this temperature gradient is an irreversible process, leading to a loss of both power and efficiency.

This effect is important in machines that use gas springs; the transient heat-transfer losses described above lead to a hysteresis in the gas springs that can be an important source of loss (Breckenridge, Heuchling, and Moore, 1971; Lee, Smith, and Faulkner, 1980; Wood, 1980). The same problem was identified by D. J. Ryden in unpublished work.

The thickness of the thermal boundary layer is $\sqrt{2\alpha/\omega}$ (Lee, Smith, and Faulkner, 1980), where α is the thermal diffusivity and ω the angular frequency. For air at 300°C and at an operating frequency of 0.65 Hz, the boundary-layer thickness is $\sqrt{[(2 \times 7 \times 10^{-5})/(2 \times \pi \times 0.65)]}$ m $\simeq$ 6 mm. A cavity or cylinder with any dimension much less than the boundary-layer thickness will behave almost isothermally. On the other hand, a much larger cavity will behave, on the whole, adiabatically. Intermediate cases will lead to more complex behavior and to higher losses.

The presently published analyses of the effect consider only conduction in an undisturbed gas in a space large enough to be considered almost adiabatic. Mixing of the gas, and scraping of the boundary layer by the moving piston or by incoming gas, are not taken into account by the theory, but their effects are accounted for by means of an experi-

Table 5.4. Comparison of Computed and Effective Schmidt Output for Adiabatic Hot Cylinder Engine.

HEATER TEMPERATURE, °C		OUTPUT POWER, W		
ACTUAL	EFFECTIVE	ACTUAL COMPUTED	EFFECTIVE SCHMIDT	DIFFERENCE
50	25	−1.27	−1.21	0.06
100	75	4.47	4.52	0.05
200	175	14.08	14.08	0.00
300	275	21.94	21.89	0.05
400	375	28.58	28.49	0.09

mentally determined enhancement factor. To the extent that the enhancement mechanism involves an effective thickening of the boundary layer, there will be a corresponding increase in the cavity dimensions or operating frequency that can be tolerated before near-isothermal behavior gives way to near-adiabatic behavior: there appear to be no published data relating to this transition. The following equation for the hysteresis loss per cycle is quoted, with a slight change of nomenclature, from Lee, Smith, and Faulkner (1980).

$$\text{Loss per cycle} = \frac{F\pi}{\sqrt{2}} \frac{\gamma}{\gamma - 1} A_s P_m \left(\frac{\Delta T}{T_m} \right)^2 \sqrt{\frac{\alpha}{\omega}} \quad (5.12)$$

where, in this and the following equations:

E_h = hysteresis loss
F = enhancement factor
k = thermal conductivity of the gas
f = frequency of operation
ω = angular frequency of operation
P_m = mean gas pressure
ΔP = pressure amplitude
T_m = mean gas temperature
ΔT = amplitude of temperature variations
V_m = mean volume
A_s = surface area
α = thermal diffusivity
γ = gas specific heat ratio

Breckenridge et al. (1971) report an F value of 2.6 for their particular gas spring. In applying the formula to small Stirling engines, Sunpower Inc., reports a value of 5 to 8 for F, with the lower values corresponding to gas spring experiments and the higher values to engines. Kangpil Lee, in discussion, has said that from his experiments, a range of 5 to 10 is typical for engines. In the absence of more definitive information relating to a particular engine configuration, an enhancement factor of 7.5 may be used—with a possible error of at least ±30 percent.

Lee, Smith, and Faulkner (1980) give an equation for the transient heat-transfer loss in a Stirling engine. However, it is fairly complex, and given the large uncertainty in the value of the enhancement factor, a simpler approach seems warranted.

For an ideal gas, the thermal diffusivity is given by

$$\alpha = k \frac{\gamma - 1}{\gamma} \frac{T_m}{P_m}$$

Inserting this into Equation (5.12) and multiplying by the frequency to give the loss per unit time, we find

$$E_h = \frac{F\sqrt{\pi}}{2} \left(\frac{\gamma}{\gamma - 1} \right)^{1/2} \sqrt{kfT_mP_m} \left(\frac{\Delta T}{T_m} \right)^2 A_s \quad (5.13)$$

It would be more convenient to evaluate Equation (5.13) in terms of the pressure amplitude—which can be obtained simply enough from the Schmidt equation—rather than the temperature amplitude. An approximate relation between temperature and pressure amplitude is the usual adiabatic one, $\gamma \Delta T/T = (\gamma - 1)\Delta P/P$. Actually, this is not strictly true for the gas in the cylinder because new gas entering from the adjacent heat exchanger does so at constant temperature. Nevertheless, this is the approximation we use, and the result of the substitution is

$$E_h = \frac{F\sqrt{\pi}}{2} \left(\frac{\gamma - 1}{\gamma} \right)^{3/2} \sqrt{kfT_mP_m} \left(\frac{\Delta P}{P} \right)^2 A_s \quad (5.14)$$

The transient heat-transfer loss increases only as the square root of the frequency and mean pressure, and is thus relatively less important in high-speed, high-pressure machines.

In the case of a typical large Fluidyne, only the hot cylinder is adiabatic and the rest of the working volume can be isothermalized. The area A_s is therefore assumed to be the mean surface area of the hot cylinder, including its top and bottom.

The gas properties are to be evaluated at the hot-cylinder temperature. If other parts of the machine, such as the connecting manifolds or the pumping arm, are also adiabatic spaces, the hysteresis loss can be calculated for them also, using Equation (5.13) or (5.14), and added to the loss from the hot cylinder.

By making some additional assumptions and approximations Equation (5.13) can be simplified still further for a large, air-filled Fluidyne operating at atmospheric pressure. Assume that the hot-cylinder diameter and stroke are equal and that the pressure compression ratio of the engine is about 2:1. The thermal conductivity of air in the range 200 to 400°C is approximately proportional to the absolute temperature with a numerical value of $k \simeq 8 \times 10^{-5} T$ $W/(m)(K)$, where T is expressed in Kelvin. Inserting this into the equation and setting $F = 7.5$ and $f = 0.75$ yields

$$E_h \simeq 0.8 \ T_e V_o^{2/3} \tag{5.15}$$

Notice that the hysteresis loss increases only as the two-thirds power of the engine displacement and therefore becomes, as expected, a smaller fraction of the output power as the engine is scaled up.

Note also, from Equation (5.14) and Table 5.5, that this loss can actually be worse with helium or hydrogen than with air, owing to their specific-heat ratio and higher conductivity; although, of course, offsetting this is the fact that the isothermal spaces behave more nearly ideally, and the heat exchangers are more effective, with a high-conductivity gas. We have ignored the transient heat-transfer losses arising from imperfectly isothermal behavior in the cold cylinder and heat exchangers, for which no analysis has been published at present.

An engine with evaporation, where the hot cylinder is largely filled with steam, would show lower transient heat-transfer losses, other

Table 5.5. Factors Relating to the Transient Heat-Transfer Loss for Various Gases at 600 K and 1 atm.

	γ	k, W/(m)(K)	$\left(\dfrac{\gamma - 1}{\gamma}\right)^{3/2} \sqrt{k}$
Air	1.40	0.047	0.033
Helium	1.67	0.271	0.132
Hydrogen	1.40	0.315	0.086
Steam	1.32	0.042	0.024

things being equal, than an air-filled engine (see Table 5.5); however, the increased pressure amplitude due to evaporation could more than offset the better thermal properties, in this respect, of steam.

The transient heat-transfer loss in the adiabatic cylinder may be the largest single factor degrading the performance. One way to reduce it is to isothermalize the hot cylinder by filling it with tubes or fins into which the liquid piston penetrates. Unless very low temperatures are used, large heat losses from evaporation and shuttle losses (see Chapter 6) will be incurred if the liquid is water. To avoid these losses, the displacer may be filled with another liquid, such as heat-transfer oil, with more favorable thermal properties. If a suitable low-density liquid, immiscible with water, could be found, it could be floated on top of the water in the hot column.

Finally, we observe from Equation (5.15) that the transient heat-transfer loss increases approximately linearly with temperature; in fact, the pressure amplitude also increases with temperature, so that the transient heat-transfer loss increases more rapidly than the temperature [Equation (5.14)]. The output power, on the other hand, increases less rapidly, as shown in Figure 5.3. For an engine with an adiabatic cylinder, therefore, we should not expect the net power output to increase indefinitely as the heater temperature is raised.

FACTORS LIMITING THE AMPLITUDE

Generally, once the hot-end temperature of a liquid feedback Fluidyne exceeds a certain level, it can begin to oscillate. Why, once this temperature is exceeded, does the amplitude of the oscillation not increase without limit?

This is not entirely a trivial question. Of course, one limit is set when the liquid piston or the float begins to hit the cylinder head at top dead center: beyond this point the liquid moves into a duct of different cross-sectional area, which modifies the resonance of the system and upsets the tuning. However, it is commonly observed that in practice, at least in machines where evaporation is suppressed, the amplitude increases steadily with increasing temperature, until one or another piston does touch the cylinder head at top dead center. What limits the oscillation at these intermediate temperatures?

If a conventional Stirling engine is allowed to run freely, increasing the temperature increases the engine speed, until frictional and heat-transfer losses absorb all the available power. This cannot happen in a Fluidyne, where the frequency is determined mainly by the resonant properties of the system and is almost independent of temperature.

Neither are the viscous losses in laminar flow or the transient heat-transfer losses necessarily strong limiters. As we have seen, both may increase approximately as the square of the amplitude and so, according to the Schmidt equation and to computations, does the indicated power. Therefore, if the power exceeds the viscous and transient heat-transfer losses at some small amplitude, it will exceed them at all larger amplitudes as well.

One important factor is probably the kinetic loss, which varies with the cube of the amplitude. At small amplitudes, it may be negligible, but as the oscillation grows it will become relatively more important and, in some cases, dominant. Figure 5.6 illustrates, for a hypothetical machine, how the sum of the losses can increase more rapidly than the

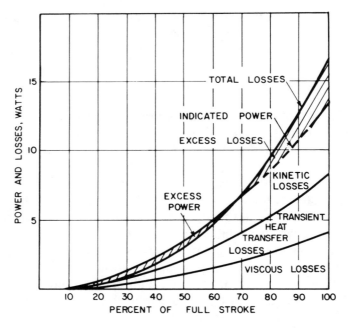

Figure 5.6. Variation of power losses and power output as amplitude increases.

indicated power as the amplitude is increased. In this particular case, the three major losses—viscous, kinetic, and transient heat transfer—are equal to each other at half full stroke. At one-tenth full stroke, the indicated output is greater than the losses by a factor of 1.5; but at 65 percent of full stroke, the losses equal the power output and this will be the limiting amplitude.

It is not only in the output column that kinetic losses may have a limiting effect. In an early experiment with a 6-in-diameter displacer tube, the regenerator was connected by a 1-in manifold—which is narrower than the minimum suggested by Equation (5.11). Only about two-thirds of the full design displacer stroke could be achieved. When the manifold diameter was increased to 2-in, this particular problem disappeared. It is worth noting that even with the 1-in-diameter manifold, the calculated kinetic gas flow losses were only about 8 percent of the gross output power; apparently the liquid feedback system could return no more than about 5 percent of the gross power to the displacer oscillation.

Another limiting effect in some machines may be poor heat transfer, from the heat source, for example. The temperature drop between the heat source and the gas will increase as the amount of heat drawn from the source increases. Thus as the amplitude of the oscillation grows, the effective gas temperature will fall and the indicated power output will therefore increase less rapidly than the square of the amplitude.

This chapter summarized the major loss mechanisms operating in the Fluidyne. Although there is, as yet, insufficient experience to arrive at general conclusions, it seems that in many cases the most significant loss will be that due to transient heat transfer. A later chapter gives an example of the design calculations for a particular machine and illustrates the relative magnitude of the various effects in a particular case. It should be stressed, however, that we have covered only the major presently known loss mechanisms; there will be other effects, so calculations of available power output are more likely to be over- than underestimates.

Chapter 6
Heat Losses

A Fluidyne operating at atmospheric pressure is a low-power-density machine and for a given cylinder size, the power output is much lower than for an internal-combustion engine or a high-pressure, high-speed Stirling engine. Consequently, heat losses by conduction through the insulation and engine components are a relatively larger fraction of the heat balance, and must be carefully minimized if high overall efficiency is to be attained. Stepdown, or shuttle, losses, present even for liquid pistons, must also be taken into account.

INSULATION AND CONDUCTION LOSSES

The calculation of heat losses through insulation is covered in many standard texts, such as Kutateladze (1963). We give only one simple example, of the heat loss from the hot engine cylinder, to illustrate the importance of good insulation.

The main heated components (hot cylinder, heat exchanger, and regenerator) are usually roughly cylindrical in shape. The heat loss, Q_i, from a long cylinder of diameter D_1 and length L surrounded by insulation of diameter D_2 is given by

$$Q_i = \frac{2\pi k L \, \Delta T}{\ln(D_2/D_1)} \tag{6.1}$$

For a 150-mm-diameter cylinder 300 mm long surrounded by 50 mm of fiber glass [$k = 0.1$ W/(m)(K)], the heat loss will be

$$Q_i = \frac{2\pi \times 0.1 \times 0.3}{\ln(24/15)} \Delta T$$
$$\simeq 0.4 \Delta T$$

If the cylinder were at 300°C and the outside of the insulation at 30°C, the heat loss from this section alone would be $0.4 \times 270 \simeq 100$ W. This could easily amount to half the total heat input of a machine of this size. Increasing the insulation thickness from 50 to 150 mm would reduce the heat loss to

$$\frac{2\pi \times 0.1 \times 0.3}{\ln(45/15)} \Delta T \simeq 50 \text{ W}$$

It is clear that to keep the heat input low, and the overall efficiency up, considerable care must be taken with the insulation of all hot sections of the engine.

Thin walls and low-conductivity material should be used to keep thermal conduction along the cylinder walls, regenerator tubes, and regenerator matrix small. This is quite feasible in low-pressure engines. These losses are addressed in Martini's *Design Manual* (1978); once again, we give only illustrative examples here.

First, consider an annular regenerator made from 1-mm-thick stainless steel cylinders. The outside diameter is 150 mm and the annular gap is 3.5 mm. The regenerator is 370 mm long (see Figure 6.1).

The area of metal is approximately $2 \times \pi \times 0.15 \times 0.001 = 9.4 \times 10^{-4}$ m^2 (9.4 cm^2). The thermal conductivity of stainless steel is 17 W/$(m)(K)$, and so if the temperature difference between the hot and cold ends of the regenerator is 300 K, the heat conducted will be

$$(9.4 \times 10^{-4}) \left(\frac{300}{0.37} \right) 17 = 13 \text{ W}$$

Area × temperature gradient × conductivity

If a higher-conductivity material were used, this loss could become very significant unless much thinner walls are used. For example, if the regenerator described here were made from aluminum alloy with a thermal conductivity of 210 W/(m)(K), the heat loss would be 160 W. This

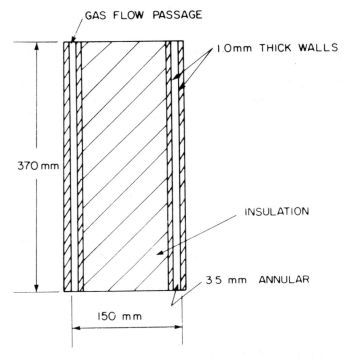

Figure 6.1. Annular gap regenerator with 1-mm-thick walls.

illustrates the need for careful choice of materials and dimensions if significant heat losses are to be avoided.

Next consider a 1-mm-thick stainless steel cylinder 150 mm in diameter and 150 mm long. Such a tube might be used as the hot cylinder of a Fluidyne. The conduction loss will be

$$\pi(0.15 \times 0.001) \left(\frac{300}{0.15} \right) 17 = 16 \text{ W}$$

SHUTTLE LOSSES WITH A CYLINDER FLOAT

This is the heat loss, sometimes known as stepdown loss, due to the motion of the displacer piston. In the stationary state, the temperature of the piston will be approximately equal to that of the adjacent cylinder wall. When the piston moves, each section of its surface moves to confront a new part of the cylinder wall, at a different temperature. Heat is transferred between the two adjacent surfaces at different temperatures.

This effect is also discussed in Martini's *Design Manual* (1978). He gives a formula for the calculation of shuttle loss which, with a change of symbols, is

$$Q_s = \frac{F(\pi/8)s^2 k \,\Delta T\, D}{Lg}$$

where s = stroke
k = thermal conductivity of gas between piston and cylinder
ΔT = temperature difference between hot and cold end of piston
D = piston diameter (or cylinder inside diameter)
L = length of piston
g = gap between piston and cylinder

F is a complicated function of the thickness and thermal diffusivity of the cylinder and piston walls. Martini notes that when the walls are thin enough for their properties to be irrelevant, F is approximately equal to 1. For thin-walled machines, this will usually be the case. In these circumstances

$$Q_s \simeq \frac{\pi s^2 k \,\Delta T\, D}{8Lg} \tag{6.2}$$

Note that this is frequency-independent. The reason is that if the frequency is, for example, increased, the time available for heat transfer during each excursion of the piston is proportionally reduced. Therefore the amount of heat transferred during each cycle is inversely proportional to the frequency. The total amount of heat transferred per unit time is equal to the amount transferred per cycle multiplied by the number of cycles. One of these factors varies inversely, and the other directly, with frequency. The overall effect is therefore independent of the frequency.

Let us calculate this effect for the typical case of a machine with a 150-mm-diameter float at the hot end, separated from a wall by 3.5

mm. The length of the float above the waterline is 250 mm and its stroke is 150 mm. The upper end of the float is at 600 K (327°C) and the lower end, in the water, is at 300 K (27°C). The thermal conductivity of air at the mean temperature of 180°C is (3.7×10^{-2}) W/(m)(K).

$$Q_s = \frac{\pi \times 0.15^2 \times 3.7 \times 10^{-2} \times 300 \times 0.15}{8 \times 0.25 \times 3.5 \times 10^{-3}} = 17 \text{ W}$$

As we shall see, this may be an underestimate. First, however, let us consider how the loss calculated in this way could be reduced. Reducing the stroke would reduce the shuttle loss rapidly, because of the square-law relation; it would also reduce the output of the machine—but less rapidly, because the output varies almost linearly with stroke if the unswept volume is reduced in proportion. Decreasing the diameter of the piston would reduce the shuttle loss linearly but decrease the power output faster, with the square of the cylinder diameter. Increasing the length of the float and the gap between the float and the cylinder would reduce the shuttle loss but would also have a bad effect on the power output of the engine: the increased dead volume would reduce the compression ratio, and the increased surface area could lead to increased transient heat-transfer losses (see Chapter 5).

To understand why the loss may be underestimated, we must understand where the heat comes from. It can come only from the heater, and thence to the float and cylinder walls, finally being shuttled down the gap between the two as already described. If the cylinder and the manifold connecting it to the heater are of thin-walled construction, and made from low-conductivity material, then little heat will be conducted directly from the heater to the hot cylinder and float. However, heat may also be transferred from the gas to the upper part of the cylinder walls, while the float is moving downward. If this happens, instead of a linear gradient all the way down the cylinder walls, the upper part may be at a high, fairly uniform temperature. The actual temperature gradient will be concentrated in a shorter length beginning somewhere below top dead center. This will reduce the effective value of L in Equation (6.2). Of course, if the gas in the cylinder were behaving truly adiabatically, i.e., with no heat transfer to its container, this effect would not take place; but we have already seen (Chapter 5) that heat transfer between the gas and the cylinder walls does take place and is indeed much enhanced compared with simple calculations of its magnitude.

There is a need for actual measurements of the shuttle losses in adiabatic hot-cylinder Fluidynes. A complex interaction between the shuttle losses and the heat loss from the gas to the cylinder walls is clearly possible, and at present good data and good theories are almost equally lacking. We know from the transient heat-transfer losses in gas springs that even in supposedly adiabatic cylinders, the heat transfer between the gas and the solid walls is higher than simple theory would predict, so simple calculations of heat loss via the gas and shuttle conduction are also likely to be incorrect. There may, however, be a way of using the experimentally determined enhancement factors for gas-to-wall heat transfer to make better estimates of the shuttle-loss effect.

SHUTTLE LOSSES WITH LIQUID PISTONS

We next consider the shuttle losses to be expected with liquid pistons. If a very low hot-end temperature is used, or if the liquid in the displacer has a high boiling point so that its vapor pressure is low at the operating temperature, the float can be dispensed with. This leads to several advantages, of which the most important may be that the hot cylinder can be isothermalized in the same way as the cold cylinder. Isothermalizing the hot cylinder will increase the ideal power output (Figure 5.3), but more importantly it can greatly reduce the transient heat-transfer losses.

With an isothermal cylinder, no separate heater is needed, so the dead volume can be reduced, thus increasing the compression ratio; alternatively the volume saved by eliminating the heater can be devoted to making a larger and more effective regenerator. Heat can be supplied to the gas in the hot cylinder through the isothermalizers directly or via the liquid (West and Geisow, 1975). In either case, the surface layers of the liquid will be close to the gas in temperature. We have already noted in the discussion of flow losses (Chapter 5) that the liquid piston in a typical Fluidyne moves as if it had a solid core lubricated by a relatively thin boundary layer. In this boundary layer, the velocity increases rapidly from zero (at the cylinder wall) to the velocity of the core. We can therefore apply Equation (6.2) directly to the liquid piston, taking g, the gap, as the thickness of the boundary layer and using the thermal conductivity of the liquid.

According to Crandall (Equation B, Appendix A, Crandall, 1927) a characteristic length in the boundary layer near an oscillating wall— the distance over which the viscous waves fall to $1/e$ of their maximum value—is $\sqrt{2\eta/\omega\rho} = \sqrt{\eta/\pi\rho f}$. We use this as a measure of the gap g between the "solid" core of the liquid and the cylinder wall. Substituting this into Equation (6.2) yields

$$Q_{s1} = \frac{\pi}{8} \frac{s^2 k \, \Delta TD}{L} \sqrt{\frac{\omega\rho}{2\eta}} \qquad (6.3)$$

This is a very simplistic derivation, and its range of applicability has not been established, although results from unpublished experiments with water under conditions typical of a Fluidyne are in good agreement with Equation (6.3). We note that according to Equation (6.3) the shuttle loss varies directly as the cylinder bore and the square of the stroke and inversely as the effective piston length, just as it does for the solid piston case discussed earlier. For the liquid piston, however, the shuttle loss is frequency-dependent, increasing as the square root of the frequency, because the effective gap between the core of the liquid and the cylinder walls is frequency-dependent.

To the shuttle losses should be added the conduction loss through the liquid:

$$Q_{sl} + Q_{cl} = \frac{\pi}{8} \frac{s^2 k \, \Delta TD}{L} \sqrt{\frac{\omega\rho}{2\eta}} + \frac{\pi D^2}{4} \frac{\Delta T}{L} k$$

$$= \frac{\pi D^2}{4} \frac{\Delta T}{L} \left(1 + \frac{s^2}{2D} \sqrt{\frac{\omega\rho}{2\eta}} \right) k$$

$$= \frac{\pi D^2}{4} \frac{\Delta T}{L} \left(1 + \frac{s^2}{2D} \sqrt{\frac{\pi\rho f}{\eta}} \right) k \qquad (6.4)$$

Before making a sample calculation from Equation (6.4), we must decide what to use as the length L over which the temperature drop takes place. One possibility would be to use the entire length of the displacer column between the hot- and cold-liquid surfaces, but this is likely to give an unrealistically low value for the temperature gradient. In the hot column the temperature, and hence the density, gradient is downward, so that the liquid is stabilized against convection currents. In the horizontal part of the displacer column, this is not so and considerable convective mixing may take place even within the core of the liquid. It would be prudent, therefore, to let L be about equal to the length of the upright hot column (which is typically one-half to one-third the length of the entire displacer column) minus half the stroke in the hot cylinder. Consider an engine with 150-mm-diameter cylinders and a stroke of 150 mm in the hot cylinder. The hot column is 0.5 m long. The hot-end temperature is 75°C and the ambient temperature is 25°C. We shall use the properties of water at the average temperature of 50°C (ρ = 990 kg/m³, η = 5.5 × 10⁻⁴ ns/m²) although a proper calculation would have to take into account the temperature dependence of the liquid properties, and especially of its viscosity. We shall also ignore heat losses due to evaporation. The frequency is 0.65 Hz.

$$Q_{sl} + Q_{cl} = \frac{\pi \times 0.15^2}{4} \frac{75 - 25}{(0.5 - 0.075)}$$

$$\times \left(1 + \frac{0.15^2}{2 \times .15} \sqrt{\frac{\pi \times 990 \times 0.65}{5.5 \times 10^{-4}}} \right) (0.64)$$

$$= 2.08(1 + 0.075 \times 1920)(0.64)$$
$$= 2.08(1 + 144)(0.64)$$
$$= 190 \text{ W}$$

This is rather a large amount of heat loss, bearing in mind that the temperature difference is only 50°C. Essentially all of it is due to the shuttle losses; the actual conduction loss is negligible.

We may not wish to use water anyway, because of its low boiling point, and so it is instructive to calculate the loss that we would have using, say, a typical engine oil [the thermal properties of engine oil may be found in Simonson (1967)].

$$Q_{sl} + Q_{cl} = 2.08 \left(1 + \frac{0.15^2}{2 \times 0.15} \sqrt{\frac{\pi \times 865 \times 0.65}{0.12}} \right) (0.14)$$

$$= 2.08(1 + 0.075 \times 121)(0.14)$$
$$= 2.08(1 + 9.1)(0.14)$$
$$= 2.9 \text{ W}$$

This is much better. However, the viscous losses will be higher than they would be for water. The viscous losses increase with the square root of viscosity; the shuttle heat losses vary inversely as the square root of viscosity. This reduction in shuttle loss is probably the reason why adding sugar to increase the viscosity of the water in the small plastic machine actually increases the available stroke (Chapter 4); at least, this was the theoretical reasoning that led to the experiment of adding sugar—and the experiment worked.

Ryden, in an unpublished work, has analyzed more rigorously the effect of periodically reversing flow on the effective thermal conductivity of a fluid in order to determine the heat loss down an oscillating column, although his calculations also ignore the temperature dependence of liquid properties.

PUMPING LOSS

If a solid float is used, the radial gap between the float and cylinder wall forms a volume that is closed at its lower end by the liquid meniscus and open to the hot cylinder at the top. As the pressure in the cylinder varies, gas flows into and out of this volume. Since the lower end of the gap is kept cold by the oscillating liquid column, extra heat must be added to this gas as it leaves the space. This loss, which presumably interacts with the shuttle loss, is generally relatively small in the Fluidyne because of the low pressure amplitude.

This topic is addressed by Martini (1978); his formula for the loss is, with a change of nomenclature and rearrangement,

$$Q_{pu} = \frac{4}{3} L \, \Delta T \left(\frac{\pi D}{k} \right)^{0.6} \left(\frac{2\Delta P f C_p}{R T_m} \right)^{1.6} g^{2.6} \qquad (6.5)$$

where L = length of float
ΔT = temperature difference between hot and cold end of float
D = diameter of float
k = thermal conductivity of gas between float and cylinder
ΔP = pressure amplitude in cylinder
f = operating frequency
C_p = specific heat at constant pressure of gas between float and cylinder

R = gas constant
T_m = mean gas temperature in gap, $(T_e + T_c)/2$
g = gap between float and cylinder

Using the same parameters and dimensions as were used in the sample calculation of shuttle loss, we find that for a gap filled with air [C_p = 1030 J/(kg)(K)] in a machine with a 2:1 pressure ratio:

$$Q_{pu} = \frac{4}{3} \times 0.2 \times 300 \times \left(\frac{\pi \times 0.15}{3.7 \times 10^{-2}} \right)^{0.6}$$

$$\times \left(\frac{2 \times 0.33 \times 10^5 \times 0.65 \times 1030}{286 \times 450} \right)^{1.6} (3.5 \times 10^{-3})^{2.6} = 1.7 \text{ W}$$

REHEAT LOSS

Reheat loss is the term given to the extra heat input needed because of the inefficiency of the regenerator. The regenerator reheats the gas as it returns to the hot cylinder, but to the extent that the regenerator is not perfectly effective, extra heat must be supplied from the heater. The topic is covered in the *Stirling Engine Design Manual* (Martini, 1978); we give a simplified treatment of the problem, suitable for approximate calculations.

The first thing we need is a measure of the heat-transfer capability of the regenerator. For this we can begin with the Nusselt number N_{nu}, a dimensionless quantity dependent on the geometry of the regenerator:

$$N_{nu} = \frac{HD_h}{k} \tag{6.6}$$

The Nusselt number is tabulated, for various standard geometries, in heat-exchanger texts and reference books. H is the heat-exchange coefficient (which is the quantity we are trying to determine), D_h is the equivalent, or "hydraulic," diameter of the regenerator, and k is the thermal conductivity of the gas. D_h is defined by

$$\frac{D_h}{L} = \frac{4A_c}{A} \tag{6.7}$$

where L = flow length of the regenerator
A_c = cross-sectional flow area
A = total heat-transfer area

Table 6.1 shows the Nusselt number for various geometries, calculated on the assumption that the surface temperature is constant [taken from Table 6.1 of Kays and London (1964)]. The table also shows the hydraulic diameter calculated from Equation (6.7) and the heat-transfer coefficient calculated by substituting the Nusselt number and the hydraulic diameter into Equation (6.6).

By way of example, to get a feel for the magnitude of the numbers involved, consider a regenerator made from the annulus, assumed to be 3.5 mm wide, between two fairly large diameter cylinders (Figure 6.1). From Table 6.1, the heat-transfer coefficient is 3.77 k/g. The working fluid is air at a mean temperature of 430 K (160 °C), which has a thermal conductivity of 3.6×10^{-2} W/(m)(K)

$$H = \frac{3.77 \times 3.6 \times 10^{-2}}{3.5 \times 10^{-3}}$$
$$= 38.8 \text{ W}/(\text{m}^2)(\text{K})$$

Similarly, a regenerator made up of 3.5-mm-diameter tubes would have a heat-transfer coefficient of

$$H = \frac{3.66 \times 3.6 \times 10^{-2}}{3.5 \times 10^{-3}}$$
$$= 37.6 \text{ W}/(\text{m}^2)(\text{K})$$

Such fine tubes would show a somewhat higher flow resistance than the annulus and may be more prone to blockage.

The heat-transfer coefficient is a measure of the performance of the regenerator per unit of surface area, and we need to calculate the performance of the regererator as a whole. According to Equation (4.97) of the *Stirling Engine Design Manual,* the performance is determined by the quantity $2/(N_{tuv} + 2)$ which represents the *in*effectiveness of the regenerator, i.e., it is a measure of the heat that the regenerator does not return to the gas. N_{tuv} is a dimensionless number, called the number

Table 6.1. Nusselt Number, Hydraulic Diameter, and Heat-Exchange Coefficient for Various Geometries.

GEOMETRY	NUSSELT NUMBER N_{nu}	HYDRAULIC DIAMETER D_h	HEAT-EXCHANGE COEFFICIENT H
Tubular regenerator, diameter d	3.66	d	$3.66k/d$
Square-tube regenerator, side b	2.89	b	$2.89k/b$
Triangular-tube regenerator, side c	2.35	$2c/3$	$3.53k/c$
Annular regenerator, gap g	7.54	$2g$	$3.77k/g$
Annular heater, gap g^*	4.86	$4g$	$1.22k/g$

*Heated from one side only.

of transfer units, representing the heat-transfer performance of the regenerator.

$$N_{tuv} = \frac{HA_r}{\dot{M}C_v} \tag{6.8}$$

The numerator is the heat-transfer coefficient per unit surface area multiplied by the surface area, i.e., the overall heat-transfer capability. The denominator is the mass flow rate multiplied by the gas-specific heat, i.e., it is proportional to the rate at which heat must be transferred to the gas. N_{tuv} is the ratio of these two quantities.

Martini, in the *Design Manual,* gives an approximate expression for the effective mass flow rate:

$$\dot{M} \simeq 3V_e f \rho_m \tag{6.9}$$

where V_e = swept volume in the hot cylinder
f = frequency of operation
ρ_m = mean gas density

As an example, let us take an air-filled Fluidyne operating at 0.6 Hz, with a swept volume in each cylinder of 1400×10^{-6} m³ (1400 cm³). The annular regenerator has a length of 370 mm and a gap of 3.5 mm and is at an effective mean temperature of 430 K (160 °C). From Equation (6.9) we first find $\dot{M}$:

$$\dot{M} \simeq 3 \times 1400 \times 10^{-6} \times 0.6 \times 0.82$$
$$= 2.1 \times 10^{-3} \text{ Kg/s}$$

The surface area of each side of the annulus is approximately $\pi \times 150 \times 10^{-3} \times 370 \times 10^{-3} = 0.174 \text{ m}^2$. The total heat transfer involves both sides of the annular gap and is therefore twice this, i.e., $A_r = 0.35$ m². The constant volume specific heat of air at 430 K is 730 J/(kg)(K) and we have already calculated that H is 38.8 W/(m²)(K). Substituting these figures into Equation (6.9) we find N_{tuv}:

$$N_{tuv} = \frac{38.8 \times 0.35}{2.1 \times 10^{-3} \times 730}$$
$$= 8.9$$

The regenerator ineffectiveness is therefore $2/(2 + 8.9) = 0.18$, or 18 percent.

A regenerator of the same length and total volume made up from 3.5-mm-diameter tubes would need a total number of 169 tubes and have a total surface area of $169 \times \pi \times 3.5 \times 10^{-3} \times 0.37 = 0.69 \text{ m}^2$. We have already shown that the heat-transfer coefficient for such a tube is 37.6 W(m²)(K), so the number of heat-transfer units N_{tuv} is given by

$$N_{tuv} = \frac{37.6 \times 0.69}{2.1 \times 10^{-3} \times 730}$$
$$\simeq 17.0$$

The regenerator ineffectiveness is therefore $2/(2 + 17.0) = 0.1$, only about half that for the annular regenerator. The price paid is greater complexity, higher flow losses, and possibly a greater tendency to regenerator blockage.

Now that we know the extent to which the regenerator falls short of thermal perfection (its ineffectiveness), it is easy to calculate the reheat loss.

Reheat loss = total heat flow through regenerator × regenerator ineffectiveness

$$Q_r = F\dot{M}C_v \, \Delta T \frac{2}{2N_{tuv}} \tag{6.10}$$

F is the effective fraction of the cycle, typically about one-third, during which gas is flowing through the regenerator toward the hot end. Using the figures calculated above for the annular and the multitubular regenerator yields

$$Q_r = \tfrac{1}{3}(2.1 \times 10^{-3})(730)(300)(0.18)$$
$$= 27.6 \text{ W (annular regenerator)}$$
$$Q_r = \tfrac{1}{3}(2.1 \times 10^{-3})(730)(300)(0.1)$$
$$= 15.3 \text{ W (multitubular regenerator)}$$

HEAT STORAGE

Although not exactly a loss mechanism, the heat that is stored in the hot components of the engine must be supplied from the heat source and can have an important effect on the overall efficiency when the engine is not run continuously, i.e., during intermittent operation.

The effect is relatively small for an adiabatic hot-cylinder machine if care is taken to use only thin-walled, low-mass construction. It would be much more significant in an isothermal hot-cylinder Fluidyne where the liquid in the hot cylinder is essentially at the hot-gas temperature.

The volume of liquid involved must be at least equal to the swept volume in the hot cylinder, and conduction down the liquid column will increase this further. Assuming that the liquid is initially at ambient temperature T_a, the amount of heat needed to reach the operating temperature is at least

$$Q_{hs} = (T_e - T_a)V_e C_l \qquad (6.11)$$

V_e is the swept volume in the hot cylinder and C_l is the specific heat of the liquid. For oil with a density of 830 kg/m³ and specific heat of 2.4×10^3 J/(Kg)(K) in a machine with a 2000×10^{-6} m³ swept volume ($V_e = V_o/\sqrt{2} = 1400 \times 10^{-6}$ m³) operating at 300°C, the initial energy necessary to raise the liquid to the operating temperature is

$$Q_{hs} = (270)(1400 \times 10^{-6})(830)(2.4 \times 10^3)$$
$$= 7.5 \times 10^5 \text{ J}$$

Such a machine might give an output of 6 W at an efficiency of 3 percent, so that the heat input in an operating condition would be 200 W. Neglecting the fact that a proportion of the stored heat can be used to prolong the running time after the heater has been switched off, we find that the total heat input for 1 h of actual operation would be

$$7.5 \times 10^5 + (200)(3600) \quad J$$
$$\underset{\text{Stored heat + input power}}{}$$
$$= 7.5 \times 10^5 + 7.2 \times 10^5 \quad J$$

In this case, the stored heat would represent half the total input, reducing the overall efficiency by half. For an 8-h run, the effect would be to reduce the overall efficiency by a factor of only 0.92. For intermittent and infrequent operation, then, the use of liquid isothermalizers in the hot cylinder could be disadvantageous: however, each specific case should be examined, because we have seen (Chapter 5) that the use of hot-end isothermalizers can increase the output of a given machine or reduce the size of machine needed for a given output. In some cases, the increased efficiency will more than pay for the heat stored in the liquid, resulting in a net gain in overall efficiency. The length of the operating period and the desired speed of response will be most important factors in determining the viability of hot-end isothermalizers.

In this chapter, we have described and calculated, at least approximately, some of more important thermal losses in a Fluidyne. In the next chapter, we shall apply these methods of calculation to a particular example.

Chapter 7
Design Example

In this chapter, we shall apply some of the loss and power calculations described in earlier sections to a hypothetical merged-cylinder liquid feedback engine (see Figure 7.1). Three points should be stressed. First, this is a hypothetical example: no machine has ever been built to these specifications, although a working Fluidyne recently constructed at the University of Calgary is based on a similar design. Second, only approximate performance calculations are given and these, as explained earlier, are more likely to overestimate than to underestimate performance. Third, there is as yet no simple way available to calculate the dynamics of the liquid columns. In assuming that good phase angles and strokes will be obtained, if the available power exceeds the losses, we are relying on the self-regulating properties of free-piston engines.

In short, the design described here is used as an example to illustrate only the mathematics described earlier; such a machine, if built, might actually show quite a different performance if, indeed, it worked at all. This partly reflects the uncertainty in much of the design data, which cannot be resolved until further experimental results with well-documented machines, especially dry machines, become available.

For this numerical example, we choose a target performance of 1.5 m^3/h to be pumped through a head of 1.5 m (roughly equivalent to 400 U.S. gal/h through a 5-ft head). This is a large enough throughput to be potentially useful for small-scale irrigation or drainage. We shall aim for an efficiency of at least 3 percent.

WORKING FLUID AND PRESSURE

The maximum efficiency reported for simple air-filled machines involving substantial water evaporation effects is less than 1 percent. To reach

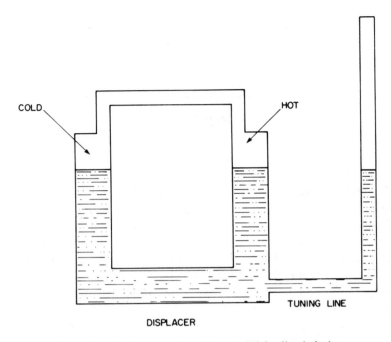

Figure 7.1. Basic merged-cylinder liquid feedback design.

our target performance, therefore, it appears that evaporation must be suppressed and the working fluid always in the gas phase. For constructional simplicity it is usual to operate with the air at a mean pressure equal to the surrounding atmosphere, which we assume to be 0.1 MPa. This can be maintained very simply by connecting the working volume to the atmosphere through a small hole or fine tube. The diameter and length of the tube, or the diameter of the hole, are chosen so that the resulting leak has no significant effect on the working cycle but is large enough to prevent the mean internal pressure from changing over a period of time.

To suppress evaporation, a float—equivalent in function to the Heylandt crown in a solid piston machine—will be placed on the liquid surface. The float can be a hollow can filled with insulation. If this is done, the displacer liquid is largely isolated from the heater and working gas, so water can be used throughout the engine.

OPERATING TEMPERATURES

High-powered, high-pressure Stirling engines typically operate with the heater head at 700 to 800°C, but special materials are needed for this. We are aiming for a simple design that can use readily available, inexpensive materials for construction, insulation, and jointing. The upper temperature is a rather arbitrary choice, but a convenient figure to use is 300 to 350°C. This temperature is within the range tolerated by many grades of fiber glass insulation, and it is also tolerated by a number of gasket-forming compounds—for example, General Electric Silicone Hi-Temp Instant Gasket, which is inexpensive and withstands 315°C. The desirability of being able easily to make and remake seals should not be underestimated.

For some purposes (e.g., use with flat-plate solar collectors) a lower expansion temperature would be preferable. An ideal Fluidyne would be able to run with a temperature difference of less than 1°C (Elrod, 1974; Geisow, 1976); however, it is known (West, 1980, 1982b) that for real machines, operation below 50 to 100°C may require the added power (and the added losses) that can be gained by allowing evaporation (Stammers, 1979). In our sample calculation, we assume a hot-gas temperature of 600 K (327°C).

A Stirling engine operating at atmospheric pressure is a low-power-density machine, so that good cooling is fairly easy to achieve. We therefore assume that the gas in the compression space is at 300 K (27°C).

OPERATING FREQUENCY

As we have seen, the frequency of operation is fairly well determined by the natural frequency of the water column in the displacer U tube in this form of Fluidyne. A displacer length of much less than 1 m is impractical for all except the smallest machines, because the minimum separation of the hot- and cold-end centerlines is equal to the cylinder bore plus the insulation thickness, and the uprights of the U tube must be long enough to accommodate the stroke easily. The natural frequency f of liquid oscillating in a U tube of length L is obtained from Equation (3.3):

$$f = \frac{1}{2\pi} \sqrt{\frac{2g}{L_D}} \quad \text{Hz} \qquad (7.1)$$

which, for values of L in the range of 1 to 2 m, corresponds to a frequency of 0.5 to 0.7 Hz. Thus, a reasonable first estimate of the operating frequency is 0.6 Hz.

SWEPT VOLUME

The next calculation is the engine-displacement, or swept, volume. A first estimate of this can be made with the aid of the Beale number (Walker, 1979), which offers a simple relation between mean cycle pressure, operating frequency, engine displacement, and power output.

$$W = B_n P f V_o \qquad (7.2)$$

Power = Beale number $\times$ mean pressure $\times$ frequency $\times$ displacement
W   bar   Hz cm^3

Beale first proposed this relation for well-developed engines operating at around 900 to 950 K (625 to 675°C), but Walker extended the concept to a wider range of conditions, and according to his figure, a typical Beale number for a machine heated to around 600 K would be 0.005.

The power needed to pump 1.5 m^3/h of water through a head of 1.5 m is $1.5 \times 10^3 \times 1.5 \times 9.81/3600 \simeq 6$ W.

We now have all the data needed to find the swept volume from Equation (7.2).

$$V_o = \frac{W_{net}}{B_n P f}$$

$$V_o = \frac{6}{(0.005 \times 1 \times 0.6)} \qquad (7.3)$$

$$= 2000 \text{ cm}^3$$

This is the total volume change during each cycle of the engine. However, in the case of an alpha-configuration machine with the hot and cold pistons moving with approximately equal strokes and about 90°

out of phase, the total volume change is $(\sqrt{2}) \times$ (the swept volume of either piston). Therefore, for our particular example, each piston should have a displacement of $2000/\sqrt{2} = 1400$ cm^3.

A circular cross-section cylinder with a stroke equal to the bore, i.e., a "square" engine, would have a diameter d such that

$$\frac{\pi d^3}{4} = 1400 \times 10^{-6} \text{ m}^3 \tag{7.4}$$

$$d = 121 \text{ mm}$$

However, it is more convenient to choose a standard size of tubing close to this—say, 150 mm (6 in)—and calculate the necessary stroke s:

$$\frac{s\pi 0.15^2}{4} = 1400 \times 10^{-6}$$

$$s = 80 \text{ mm}$$

DEAD VOLUME

The optimum compression ratio in a Stirling machine is about 2:1 (Rallis and Urieli, 1976). For an alpha-configuration machine, if both pistons have a swept volume V and differ in phase by $90°$, the total volume of the working gas in the cylinders, excluding clearance spaces, is given by

$$\frac{V}{2}(1 + \cos \omega t) + \frac{V}{2}(1 + \sin \omega t)$$

$$= V\left(1 + \frac{1}{\sqrt{2}} \cos\left(\omega t - \frac{\pi}{4}\right)\right) \tag{7.5}$$

If the unswept volume available for clearance space, heater, and regenerator is V_D, then the volume compression ratio, i.e., the ratio of maximum to minimum volume is

$$\frac{V(1 + 1/\sqrt{2}) + V_D}{V(1 - 1/\sqrt{2}) + V_D} \tag{7.6}$$

which is equal to 2 when $V_D = 1.12$ V. The 1500 L/h machine has a cylinder swept volume of 1400 cm^3, so the unswept volume available is $1400 \times 1.12 = 1570$ cm^3.

PUMP VOLUME

At a frequency of 0.6 Hz, the number of strokes per hour is $0.6 \times 3600 = 2160$. To pump 1.5 m^3/h therefore requires $1.5/2160 = 700 \times 10^{-6}$ m^3 = 700 cm^3 per stroke. The pumping system can be gas-coupled to the Fluidyne by means of an air-filled pipe (see Figure 7.2). The mean volume of air in this pipe must be at least 350×10^{-6} m^3 (350 cm^3),

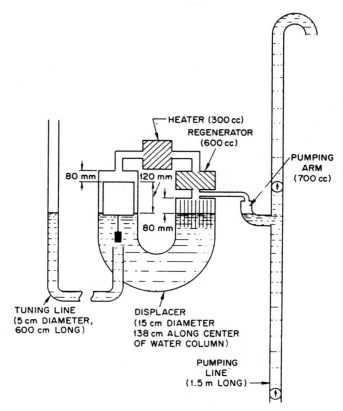

Figure 7.2. Conceptual layout and major dimensions for sample design.

rising to 700×10^{-6} m^3 (700 cm^3) and falling to zero during each stroke. The remaining unswept volume, $(1570 - 350) \times 10^{-6} = 1220 \times 10^{-6}$ m^3 (1220 cm^3), is available for heat exchangers, regenerator, clearances, and connecting pipes. As a first approximation, we allocate 900×10^{-6} m^3 (900 cm) of this to heat exchange and the remainder (approximately 300 cm^3) to clearances and connections.

HEAT-EXCHANGE REQUIREMENTS

No separate cooler is needed because the cold cylinder can be subdivided into small, nearly isothermal cavities by means of tubes (Walker has suggested drinking straws), strips, Aeroweb or other material.

In the hot cylinder of the machine, the working gas can be separated from the liquid piston to prevent heat loss and evaporation. A float may be used for this purpose (West and Geisow, 1975). If a liquid with suitable physical properties could be used in the displacer, or at least at the hot end of the displacer, the float would be unnecessary. However, as long as a float is used, the hot cylinder cannot easily be isothermalized, so a separate heater will be needed (see Figure 7.2 for the conceptual layout and major dimensions).

How should we divide up the available volume between the heater and the regenerator? Equations (2.2) and (3.2) of the report by Rallis and Urieli (1976) indicate the effect of regenerator effectiveness on the heat input required for a given power output. Using either of these equations we find that the heat input required with no regeneration is about three times that required with perfect regeneration (for a compression ratio of 2:1, a temperature ratio of 2:1, and a gas specific-heat ratio of 1.4). This implies that under these conditions, during each cycle a perfect regenerator returns about twice as much heat as is supplied by the heater. Therefore, as a rough approximation, we can estimate that the area available for heat exchange in the regenerator should be about twice the surface area available inside the heater. As an even more approximate guess, we can allocate about twice as much volume to the regenerator as to the heater.

For the example calculation that we are following, this leads to about 600×10^{-6} m^3 (600 cm^3) for the regenerator and 300×10^{-6} m^3 (300 cm^3) for the heater.

TUNING-LINE LENGTH AND DIAMETER

We calculate the tuning-line length from Equation (3.26). To do this, we need to know how the unallocated unswept volume (300 cm^3) is divided between isothermal and adiabatic spaces. The question is not crucial, since this is a small term; we will simply assume that it is divided into approximately the same proportion as that of the rest of the gas, i.e., about 2:1. Table 7.1 indicates the mean gas distribution.

Substitution of these figures and a displacer diameter of 150 mm into Equation (3.25) yields the following relation between L_t and R_t for an operating frequency of 0.6 Hz:

$$L_t = \frac{\dfrac{\pi \times R_t^2 \times 10^5}{10^3(1800 \times 10^{-6} + 1150 \times 10^{-6}/1.4)} + 9.81\left(1 + \dfrac{R_t^2}{2 \times 0.075^2}\right)}{4\pi^2 \times 0.6^2}$$

$$\tag{7.7}$$

$$= (8.49 \times 10^3)R_t^2 + 0.69 \tag{7.8}$$

$$= (2.12 \times 10^3)D_t^2 + 0.69$$

Table 7.2 shows the length of the tuning line and the volume of liquid in the tuning column for a useful range of tuning-line diameters, using the parameters of our sample design.

We have seen that if the machine is to operate in a resonant mode, the length of the tuning line increases approximately as the square of its diameter. On the grounds of compactness and material economy a narrow diameter and short length would be the best choice. However, other important factors intervene.

Table 7.1. Mean Volume of Adiabatic and Isothermal Spaces (10^{-6} m^3).

	ISOTHERMAL	ADIABATIC
Hot cylinder	. . . .	700
Cold cylinder	700	. . . .
Regenerator	600	. . . .
Heater	300	. . . .
Pump	. . . .	350
Other	200	100
Total	1800	1150

Table 7.2. Tuning-Line Diameter,
Length, and Volume.

DIAMETER, mm	LENGTH, m	VOLUME, L
25.0	2.0	1.0
37.5	3.7	4.1
50.0	6.0	11.8
62.5	9.0	27.6

First, the narrower the tube, the greater the amplitude of the motion. At high amplitudes, when the peak downward acceleration exceeds $1g$, the water surface becomes unstable, and some of the water is "left behind." Consequently, the effective volume of moving water during the downstroke of the open end is lower than that during the upstroke. For a machine operating at 0.6 Hz, this will happen when the amplitude of movement in the tuning column exceeds $9.81/(4\pi^2 \times 0.6^2)$, or about 0.7 m. For our example machine, the amplitude will be greater than 0.7 m if the diameter of the tuning column is less than $2\sqrt{2000 \times 10^{-6}/(2\pi \times 0.7)} = 43$ mm.

Second, unless the stroke in the tuning column is very much less than its mean volume, the moving mass will vary considerably during the cycle, thus invalidating our assumption of simple harmonic motion.

The example machine has a stroke of 2000 cm³. This is more than the total volume in the tuning line when its diameter is 25 mm but falls to less than 20 percent (i.e., the volume varies from the mean by less than ± 10 percent during the stroke) when the tuning line is 50 mm or more in diameter. This is, perhaps, a more comfortable figure.

Third is the question of flow losses in the tuning line. These can be estimated using Equations (5.4) and (5.7) and either substituting the relation between R_t and L_t from Equation (7.8) or using the numerical relation given in Table 7.2.

From Equation (5.4):

$$E_v = \frac{\pi\sqrt{\pi\rho f\eta}\ L_t f^2 V_o^2}{R_t^3}$$

$$= \frac{\pi\sqrt{\pi\rho f\eta}\ (8.49 \times 10^3 R_t^2 + 0.69)f^2 V_o^2}{R_t^3} \tag{7.9}$$

$$= (\pi\sqrt{\pi} \times 10^3 \times 0.6 \times 10^{-3})(8.49 \times 10^3 R_t^2 + 0.69) \times$$
$$(0.6^2)(2000 \times 10^{-6})^2/R_t^3$$
$$= \frac{5.27 \times 10^{-2}}{R_t} + \frac{4.28 \times 10^{-6}}{R_t^3}$$
$$= \frac{0.105}{D_t} + \frac{3.43 \times 10^{-5}}{D_t^3} \tag{7.10}$$

In calculating E_k for the example machine, we assume that the tuning line has two 90° smooth bends ($K = 0.2$) and a conical enlargement ($K = 0.2$) at the displacer end. Therefore $\Sigma K = 0.2 + 0.2 + 0.2 = 0.6$.

From Equation (5.7):

$$E_k = \frac{0.42\Sigma K\pi\rho f^3 V_o^3}{2R_t^4} \tag{7.11}$$
$$= \frac{(0.42 \times 0.6 \times \pi \times 10^3 \times 0.6^3)(2000 \times 10^{-6})^3}{2R_t^4}$$
$$= \frac{6.84 \times 10^{-7}}{R_t^4} \tag{7.12}$$
$$= \frac{1.09 \times 10^{-5}}{D_t^4} \tag{7.13}$$

Table 7.3 shows the estimated liquid flow losses, calculated from Equations (7.10) and (7.13), in the tuning line of the example machine.

The ideal power output according to the Schmidt equation is only 24 W, which is reduced to 23 W, according to the ALPHA WEST computer program (West, 1979) when allowance is made for the adiabatic hot cylinder. We are probably unwilling to devote more than, say, 20

Table 7.3. Liquid Flow Losses in the Tuning Line.

DIAMETER, mm	VISCOUS LOSS, W	KINETIC LOSS, W	TOTAL, W
25.0	6.4	27.2	33.6
27.5	3.4	5.5	8.9
50.0	2.4	1.7	4.1
62.5	1.8	0.7	2.5

percent of this to the flow losses, corresponding to a mechanical efficiency of 80 percent. The minimum acceptable tuning-line diameter is therefore about 50 mm. However, if Equation (5.4) seriously underestimates the viscous losses, a larger diameter tube will be called for (see Chapter 5).

Note that the kinetic losses for a 50-mm-diameter tube are almost half the total flow losses calculated as above. This emphasizes the need for better design data concerning minor pipe losses in oscillating flow.

DISPLACER LENGTH

The actual length L_D of the displacer water column may be calculated from Equation (3.3), (3.7), or (3.14) as appropriate. If the displacer is a simple U tube of constant cross section, the applicable formula is Equation (3.3) which may be rearranged:

$$
\begin{aligned}
L_D &= \frac{g}{2\pi^2 f^2} \\
&= \frac{9.81}{2\pi^2 \times 0.6^2} \\
&= 1.38 \text{ m}
\end{aligned}
\tag{7.14}
$$

INSULATING FLOAT

The main object of the insulating float is to prevent evaporation in the hot chamber and heat loss down the hot-liquid column. The wall of the hot cylinder above the float will be heated by thermal conduction from the hot working gas and the wall alongside the float will be heated by shuttle losses. Water coming into contact with walls during a later phase of the cycle will carry off much of this heat. Therefore if the float is to be fully effective, its length above the waterline should be longer than the stroke in the hot cylinder. If the float is rigid (probably a hollow cylinder or a lightweight foamed material), the clearance gap between the float and the cylinder wall is part of the unswept volume of the engine, which should be minimized. This is another compromise. A float length above the water level equal to $1\frac{1}{2}$ times the stroke might be reasonable. For the example machine, this is 120 mm.

One more parameter to be decided upon is the annular gap between the float and the cylinder wall. If it is too wide, the unswept volume will be very high, thus reducing the compression ratio. If it is too narrow, the construction tolerances will be difficult and expensive to achieve, throwing away some of the simplicity we are seeking. For the example machine, a gap of $3\frac{1}{2}$ mm (slightly more than $\frac{1}{8}$ in), corresponding to an annular volume of 195 cm^3, should not be too difficult to achieve; it is within the allowance already made for isothermal unswept volume.

If necessary, the float is weighted so that it is stable in an upright position. A liquid insulating float could be a good solution if a suitable liquid can be identified. A liquid of low vapor pressure and density, thermally stable and immiscible with water even over long periods, could be floated on top of the hot-liquid column. The thermal properties required of such a liquid were discussed in Chapter 5. No liquid with the appropriate parameters and the necessary stability has yet been identified: experiments with heat-transfer oil have been successful over relatively short periods, but after several days of continuous operation separation of the components of the oil, and mixing with the water, was observed.

If an inexpensive liquid with low enough viscosity and appropriate thermal properties could be found, it could be used throughout the displacer and tuning line so that no water would be involved in the engine section at all. Either of these methods for using a low-vapor-pressure liquid would do away with the need for a solid float altogether.

If the hot cylinder were truly adiabatic, with no heat transfer between the working gas and the cylinder walls and liquid surface, then no float would be needed and any convenient liquid, including water, could be used in the cylinder. We have seen that there is heat transfer in the cylinder, leading to the transient heat-transfer losses, but that this is ideally confined to a relatively thin boundary layer close to the walls, cylinder head, and liquid surface; evaporation would be expected to increase the thickness of this boundary layer, but not by a very large factor, because the diffusion coefficient for heat and for water vapor in air are fairly similar. Consequently, in a large enough cylinder, heat transfer from the working gas will be negligible compared with volume effects, and no float will be needed. Experiments in a 4-in-diameter cylinder showed that a float similar to the one in Figure 7.2 did improve performance significantly. However, it is suspected that even a very short float might have worked almost as well (West, 1977). More experimental data, including data from larger machines, are needed to resolve this issue.

SUMMARY OF DESIGN SIZES

The main parameters of the design examples are given in Table 7.4. We shall use these dimensions to calculate other performance figures relating to the example.

IDEAL POWER OUTPUT

If both cylinders of the engine were isothermal, we could calculate the power output easily from the Schmidt equation as 23.9 W. By means of a numerical integration (e.g., using the ALPHA WEST program) this can be corrected for the effect of the adiabatic hot cylinder. The ideal power output of the adiabatic hot cylinder engine is 22.6 W at a heater temperature of 600 K (327 °C) and with a 90° phase angle between the hot and cold cylinders.

Figure 7.3 shows the variation of the output of the adiabatic hot-cylinder machine with phase angle and with temperature. The computed ideal output power varies only slowly with phase angle—at a

Table 7.4.   Cylinder, Float, Heat Exchanger, and Tuning Column Size.

Cold-cylinder stroke	79 mm
Cold-cylinder diameter	150 mm (isothermalized)
Hot-cylinder stroke	79 mm
Hot-cylinder diameter	150 mm (adiabatic)
Hot-cylinder float length	120 mm (above water level)
Float-cylinder gap	3.5 mm
Tuning-column diameter	50 mm (alternative design: 62.5 mm)
Tuning-column length	6.0 m (alternative design: 9.0 m)
Pump-arm mean volume	350 cm^3 (half the pumping stroke)
Regenerator internal volume	600 cm^3
Heater internal volume	300 cm^3
Manifold and dead-space volume	200 cm^3 (isothermal)
	100 cm^3 (adiabatic)
Heater temperature	327 °C
Cold-cylinder temperature	27 °C

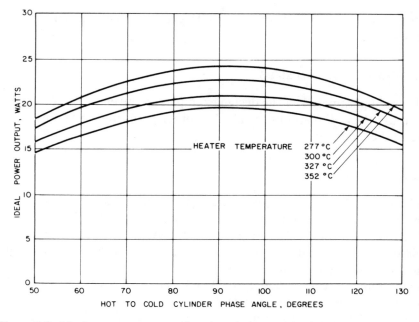

Figure 7.3. Ideal power output as a function of phase angle and heater temperature.

120° phase angle, the ideal output is still only about 10 percent below its maximum value—and this has important consequences, because the major sources of mechanical power loss are flow losses and transient heat-transfer loss, both of which decrease rapidly as the phase angle is increased. Therefore, if the losses have been underestimated, the machine may still run, but at a phase angle larger than 90°. This effect has been observed in practice.

FLOW LOSSES

We have already calculated the flow losses at the nominal operating point—Table 7.3—but it is worthwhile calculating the effect of varying the hot- to cold-piston phase angle on the flow losses.

Assuming that both cylinders operate at full stroke ($V_e = V_c$) we find that the stroke in the tuning line V_t will be

$$V_t = \sqrt{V_e^2 + V_c^2 + 2V_eV_c \cos\theta}$$

Therefore, if $V_e = V_c$ the relation between V_t for a 90° phase and any other phase angle θ is

$$\frac{V_{t,\theta}}{V_{t,90}} = (1 + \cos\theta)^{1/2} \qquad (7.15)$$

At a fixed frequency, the viscous loss varies with the square of the displacement and the kinetic loss with the cube of the displacement. Equation (7.15) gives the variation of the displacement with phase angle and, when combined with the losses already calculated for a 50-mm-diameter tube, can be used to calculate the losses as a function of phase angle.

At this point, we may pause to calculate the Reynolds number for the water flow in the tuning column. Actually, for the case of the 90° phase angle, the calculation has already been done, with the results shown in Table 7.1. At a phase angle of 90°, the flow in a 50-mm-diameter output column has a Reynolds number of 95,000 and is expected to be turbulent, since the critical Reynolds number is only 67,000. Consequently, the viscous losses may be considerably higher than those indicated in Table 7.5. However, as the phase angle between the two cylinders widens, the output swept volume falls and so does the flow

Table 7.5. Flow Loss Versus Cylinder Phase Angle for a 50-mm-Diameter Tuning Line.

PHASE ANGLE, °	SWEPT VOLUME, cm³	VISCOUS LOSS, W	KINETIC LOSS, W	TOTAL LOSS, W
50	2563	3.95	3.58	7.5
60	2449	3.60	3.12	6.7
70	2317	3.22	2.64	5.9
80	2167	2.82	2.16	5.0
90	2000	2.40	1.70	4.1
100	1818	1.98	1.28	3.3
110	1622	1.58	0.91	2.5
120	1414	1.20	0.60	1.8
130	1195	0.86	0.36	1.2

velocity in the output column. Table 7.5 shows that at a phase angle of 120°, the swept volume, and hence the flow velocity, is 30 percent below its value at 90°, and this is sufficient to reduce the Reynolds number from 95,000 to 67,000 so that the flow should be mainly laminar. If smaller phase angles are used, then a larger diameter output tube will be needed; Table 5.1 indicates that the flow in a 62-mm-diameter tube should be nonturbulent even at a phase angle of 90°. Such a tube must, however, be 50 percent longer than the 50-mm-diameter one (see Table 7.2).

TRANSIENT HEAT-TRANSFER LOSS

The most important mechanical loss remaining to be calculated is the transient heat-transfer loss. The approximate equation is Equation (5.14):

$$E_H = \frac{F\sqrt{\pi}}{2} \left(\frac{\gamma - 1}{\gamma} \right)^{3/2} \sqrt{Kf T_m P_m} \left(\frac{\Delta P}{P_m} \right)^2 A_s$$

To evaluate this, we must know the amplitude of the pressure variations in the engine. This may be calculated, approximately, from the usual Schmidt equation. According to the Schmidt equation [see, e.g., Walker (1973)]:

$$\frac{P_{max}}{P_{min}} = \frac{1 + \delta}{1 - \delta} \tag{7.16}$$

and
$$P_{mean} = P_{max} \left(\frac{1 - \delta}{1 + \delta} \right)^{1/2} \tag{7.17}$$

where P_{max} = maximum cycle pressure
P_{mean} = mean cycle pressure
P_{min} = minimum cycle pressure
ΔP = pressure amplitude
θ = hot- to cold-piston phase angle
$\delta = (\tau^2 + \kappa^2 + 2\tau\kappa \cos \theta)^{1/2}/(\tau + \kappa + 2S)$
$S = 2 X \tau/(1 + \tau)$, the reduced dead volume ratio

$X = V_D/V_e$, the actual dead volume ratio
$\tau = T_c/T_e$, the temperature ratio
$\kappa = V_c/V_e$, the ratio of cold- to hot-cylinder swept volume

The pressure amplitude ΔP can be calculated from Equation (7.16):

$$\Delta P = \frac{(P_{max} - P_{min})}{2} = P_{max} \frac{\delta}{1 + \delta} \tag{7.18}$$

and the ratio $\Delta P/P_{mean}$ is

$$\begin{aligned}
\frac{\Delta P}{P_{mean}} &= \frac{\Delta P}{P_{max}} \frac{P_{max}}{P_{mean}} \\
&= \frac{\delta}{1 + \delta} \frac{(1 + \delta)^{1/2}}{(1 - \delta)^{1/2}} \\
&= \frac{\delta}{(1 - \delta^2)^{1/2}}
\end{aligned}$$

and the square of this ratio, which is the quantity appearing in the expression for transient heat-transfer loss, is

$$\left(\frac{\Delta P}{P_{mean}} \right)^2 = \frac{\delta^2}{1 - \delta^2} \tag{7.19}$$

We have designed this machine to have a volume compression ratio of 2:1, and the ratio V_D/V_e is therefore equal to 1.12:1 [see Equation (7.6)]. Also, the two cylinders have the same swept volume at full stroke, i.e., $\kappa = 1$. Inserting these figures into the expression for δ yields

$$\delta = \frac{(1 + \tau^2 + 2\tau \cos \theta)^{1/2}}{1 + \tau + 4.48 \, \tau/(1 + \tau)} \tag{7.20}$$

Using Equation (7.20) we can calculate the square of the pressure amplitude ratio for the example machine at various hot-end temperatures and phase angles. The results are shown in Table 7.6.

As we expect, the pressure amplitude does not vary rapidly with

Table 7.6. $(\Delta P/P_{mean})^2$ at Various Expansion Temperatures and Phase Angles.

EXPANSION TEMPERATURE, °C	PHASE ANGLE, °								
	50	60	70	80	90	100	110	120	130
277	0.257	0.232	0.206	0.179	0.153	0.128	0.104	0.083	0.065
300	0.262	0.237	0.211	0.184	0.157	0.132	0.108	0.087	0.068
327	0.268	0.243	0.216	0.189	0.162	0.137	0.113	0.091	0.073
352	0.274	0.248	0.221	0.194	0.167	0.141	0.117	0.096	0.077

expansion-space temperature in this range, but it does vary considerably with phase angle.

These values of pressure ratio can be used to calculate the transient heat-transfer loss. Table 7.7 shows the results for the following values of the machine parameters.

$F = 7.5$	Enhancement factor
$\gamma = 1.4$	Gas specific heat ratio
$k = 4.6 \times 10^{-2}$ W/(m)(K)	Gas thermal conductivity
f = 0.6 Hz	Frequency
$T_e, T_m = 600$ K	Hot gas temperature
$P_m = 10^5$ Pa	Mean gas pressure
$A_s = 5.4 \times 10^{-2}$ m^2	Mean surface area in hot cylinder

NET AVAILABLE POWER

Figure 7.4 shows the losses already calculated (tuning-line flow losses and transient heat-transfer losses) as a function of phase angle for a hot-gas temperature of 600 K (327°C). The same graph shows the power output for an ideal adiabatic hot-cylinder engine. The difference between the two is the net power available for overcoming other losses and for pumping. The power, net of tuning line and transient heat-transfer losses, is shown in Figure 7.5 as a function of phase angle and

Table 7.7. Transient Heat-Transfer Loss at Various Phase Angles; Hot Gas Temperature, 600 K.

Phase angle, °	50	60	70	80	90	100	110	120	130
Transient heat-transfer loss, W	18.9	17.1	15.2	13.3	11.4	9.7	8.0	6.4	5.1

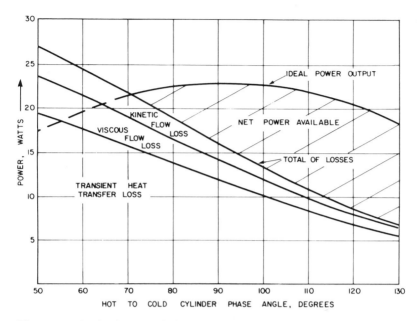

Figure 7.4. Major losses and ideal power output as a function of phase angle.

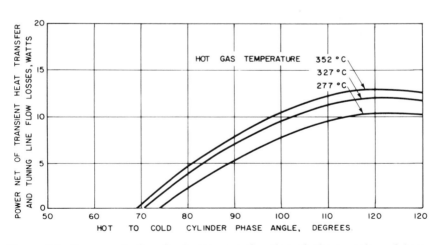

Figure 7.5. Power, net of major losses, as a function of phase angle and heater temperature.

hot-end temperature. Remember that the viscous flow losses may be substantially underestimated, especially at phase angles below 120°, unless a wider output tube is used.

Figures 7.4 and 7.5 reveal a number of interesting insights into the operation of the Fluidyne—if the theoretical data that have gone into them can be believed. First, for a machine of this design, transient heat transfer in the adiabatic spaces is by far the largest single source of loss, at least when the tuning-line flow behavior is laminar. This emphasizes the need for better methods of calculating the transient heat-transfer loss in a Fluidyne, and also the importance of minimizing this source of loss; all possible care should be taken to isothermalize as much as possible of the working volume, including the pumping arm, and if possible the hot cylinder should be isothermalized also if maximum output is the aim.

Second, there may in practice be advantages to operating with a phase angle considerably greater than 90°, because the most important losses (tuning-line flow losses and transient heat-transfer losses) decrease much more rapidly than does gross power output as the phase angle is increased—at least up to 120 or 130°.

Table 7.8 summarizes the power output and loss calculations for one particular case, a phase angle of 120° and a hot-end temperature of 600 K (327°C).

These calculations do not take into account the interaction of the various losses with each other. Nor do they take into account the loading placed on the system by the pump. At present, there is no simple way to calculate the effect of the pump on the system, but some physical arguments indicate what may happen. Water will begin to move in the pumping arms only when the pressure in the working gas exceeds the threshold necessary to open the valves against the head of water.

Table 7.8. Summary of Typical Power Output and Major Loss Results.

QUANTITY	POWER LOSS, W	POWER AVAILABLE, W
Schmidt output	. . .	20.7
Effect of adiabatic hot cylinder	0.6	20.1
Tuning-line viscous loss	1.2	18.9
Tuning-line kinetic loss	0.6	18.3
Transient heat-transfer loss in hot cylinder	6.4	11.9
Available for other losses and output	. . .	11.9

Table 7.9. Range of Uncertainty on Typical Power and Loss
Calculations.

QUANTITY	RANGE OF VALUES, W
Computed ideal output	20.1
Tuning-line viscous loss	1.2–4.5
Tuning-line kinetic loss	0.6–0.6*
Transient heat-transfer loss in hot cylinder	4.5–9.0
Available for other losses and output	6.0–13.8

*No information available on the accuracy of the formulas for kinetic losses.

Beyond this point, the pump will draw power from the varying gas pressure and in so doing will reduce the amplitude of the pressure variation in the engine below what it would otherwise be. This will reduce the gross power output and also the transient heat-transfer losses. The volume change in the pumping arm will lag in phase behind the pressure variation in the engine: however, it will not be sinusoidal (usually it will not even be continuous, because there will be portions of the period when both pump valves are closed—see appendix.)

Table 7.8 does not reflect the fact that there is great uncertainty about the accuracy and applicability of the formulas used to make the loss calculations. Table 7.9 attempts to indicate roughly the range of uncertainty; the ranges shown are based on the different experimental results on oscillating flow losses and on the hysteresis loss enhancement factors that were discussed in Chapter 6. The final margin of uncertainty is rather wide—and may still be underestimated.

MAJOR HEAT LOSSES

The ideal thermodynamic heat input with a hot-end temperature of 600 K (327 °C) and a phase angle of 120° is approximately 45 W, according to the ALPHA WEST program. To this must be added the heat loss through the insulation, the shuttle loss, and the reheat loss.

Let us suppose that the regenerator and heater are both made in annular form, with the same diameter (150 mm) as the cylinders (Figure 7.6). We set the annular gap at 3.5 mm, the same as the gap around the float, since this should not pose too difficult a construction problem. However, it should be remarked that it is very important in simple annular regenerators to ensure that the gap be uniform; otherwise the

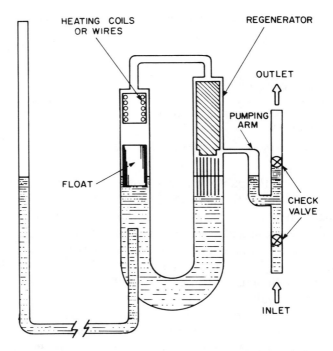

Figure 7.6. Overall layout of sample design showing annular regenerator and heater.

gas flow will be preferentially in the wider part of the gap, leading to an increased flow rate and much reduced heat-transfer efficiency in that region.

The volume of the heater is 300×10^{-6} m^3 (300 cm^3), so its length will be $(300 \times 10^{-6})/[\pi(150 \times 10^{-3})(3.5 \times 10^{-3})] = 185$ mm. Similarly, the regenerator will be 370 mm long; for the purposes of calculating heat loss through the insulation, we assume that the regenerator

Table 7.10. Major Heat Loads, in Watts.

Thermodynamic input	45
Insulation loss	75
Reheat loss	28
Regenerator matrix conduction	13
Shuttle loss	12
Total	173

is at the mean temperature of the engine, i.e., $(600 + 300)/2 = 450$ K, or $177°C$. If the insulation is 150 mm of fiber glass with a thermal conductivity of 0.1 W/(m)(K), then from Equation (6.1) the heat loss will be

$$Q_i = \underbrace{\frac{2\pi \times 0.1 \times 0.08 \times 300}{\ln(0.45/0.15)}}_{\text{Hot cylinder}} + \underbrace{\frac{2\pi \times 0.1 \times 0.185 \times 300}{\ln(0.45/0.15)}}_{\text{Heater}}$$

$$+ \underbrace{\frac{2\pi \times 0.1 \times 0.37 \times 150}{\ln(0.45/0.15)}}_{\text{Regenerator}}$$

$$= 13.7 + 31.7 + 31.7 \simeq 75 \text{ W}$$

We have already calculated, in Chapter 6, the heat loss down the regenerator walls as 13 W.

The shuttle loss is calculated according to Equation (6.2):

$$Q_s = \frac{\pi s^2 k \, \Delta T \, D}{8Lg}$$

where s = stroke = 80 mm
k = thermal conductivity of air = 4.6×10^{-2} W/(m)(K)
ΔT = temperature difference = 300 K
D = float diameter = 150 mm
L = float length = 120 mm
g = annular gap, round float = 3.5 mm

$$Q_s = \frac{\pi \times 0.08^2 \times 4.6 \times 10^{-2} \times 300 \times 0.15}{8 \times 0.12 \times 3.5 \times 10^{-3}}$$

$$= 12 \text{ W}$$

As explained in Chapter 6, this is probably an underestimate.

The reheat loss was calculated in Chapter 6 for this case: it is approximately 28 W.

The total heat input to make up the thermodynamic input and the major heat losses is given in Table 7.10.

If the design output of 6 W (pumping power) is achieved, the overall efficiency will be very approximately $6/173 \simeq 3\frac{1}{2}$ percent.

In this chapter, we have outlined the major elements of a basic design calculation. Such a calculation indicates roughly the relative importance of some of the effects known to take place in a Fluidyne. Neither the theory nor the existing experimental data are good enough to give great confidence in the exact quantitative results derived in this chapter. Nevertheless, in identifying major power and heat loss mechanisms, we can at least get a broad idea of the likely performance, and the likely reasons for poor performance, of an actual engine.

Chapter 8
Effect of Evaporation on Efficiency and Output

Many of the Fluidyne engines that have been built, including those illustrated in Chapter 4, allow water to evaporate from the hot cylinder. It is known, in a general way, that this evaporation increases the power output for a given size of machine but usually increases the heat input by a still larger factor. The efficiency is thereby reduced. However, there are very important practical exceptions to this: small machines may have such high flow and hysteresis losses that they will not run at all as dry machines. If the extra power available from evaporation is sufficient to overcome these losses, the engine will run and therefore have a finite efficiency that, however low, is at least greater than that of the inoperable dry machine. Similarly, with a small temperature difference between the hot and cold cylinders, the losses become a larger fraction of the output even in large machines, and the extra power derived from evaporation may be needed to allow the engine to run.

IDEAL POWER OUTPUT AND POWER INPUT FOR WET AND DRY MACHINES

The theory of the wet cycle Fluidyne is very far from complete, and no more than rough estimates of performance can be made (West, 1982b). Calculations based on a greatly simplified model (Walker and Agbi, 1973a, b) show that a two- to threefold increase in indicated specific power may be achieved by the use of a two-component, two-phase mixture of air and water. It appears that still larger increases may be possible through the use of a single-component, two-phase working fluid, but no results (experimental or theoretical) seem to have been published on such a system, which would require the rigorous exclusion of air from the working space and use of vapor only for the working fluid. We there-fore rely on the Walker and Agbi calculations and consider only air–vapor mixtures. In fact, we shall consider only mixtures of air and water vapor.

The comparison between wet and dry machines will involve three hypothetical engines, with cylinder diameters of 150, 50, and 16 mm. These dimensions correspond very roughly to, respectively, the design given in Chapter 7, although with a longer stroke, and to the laboratory prototype pump described by West and Pandey (1981); the machine shown in Figure 4.5 and the one described by Reader et al. (1981); and the machine shown in Figure 4.6. In each case, a stroke-to-bore ratio of 1 and a phase angle of 120° will be assumed. A hot-end temperature of 327° will be assumed for the dry case, and 90°C for wet operation. Note that the water temperature found in wet machines cannot be much higher than this if (as is usually the case) no boiling takes place even during the downstroke of pressure.

The ideal specific output shown in Table 8.1 is calculated from the following simple formula: (power/cycle)/(unit swept volume) $\simeq$ (mean pressure $\times$ temperature difference)/(temperature sum) (see West, 1981). A factor of $2\frac{1}{2}$ improvement, as suggested by the Walker and Agbi results, is used to calculate the effect of evaporation on the specific output. The ideal heat input for the dry machine is calculated from the Carnot efficiency; and for the wet machine it is calculated as the latent heat of vaporization of 1 L of water vapor at atmospheric pressure. Efficiency is simply output divided by input.

Apparently, the basic efficiency of the wet machine is, as we should expect, lower than a dry Fluidyne; but what about the effect of losses?

Table 8.1. Ideal Machines.

IDEAL PERFORMANCE	DRY		Wet
	600 K*	363 K	363 K
(Output/cycle)/L, Joules	33	9.5	24
(Input/cycle)/L, Joules	66	55	1400
Efficiency (percent)	50†	17†	1.7

*600 K and 363 K are effective hot-space temperatures. Cold space tem-perature is 300 K (27°C).
†Carnot efficiency.

SOME IMPORTANT POWER LOSSES

The three most important of the known losses in the Fluidyne, which were discussed in Chapter 5, are flow losses, hysteresis losses, and the loss due to the irreversible mixing of gas from the heater with the gas, at a different temperature, in the expansion cylinder, called the adiabatic loss. Approximate calculations of these effects, made with the equations given in Chapter 5, are shown in Table 8.2 for the three different sizes and two different operating temperatures under consideration.

Table 8.2 is very revealing if the numbers in it can be believed. A positive net output means that the machine will be able to run under the conditions cited.

At low temperatures (90°C) only the wet machines show a positive output and the dry machines are therefore not expected to operate successfully at this temperature. A major reason for this is the high value of the hysteresis loss, and a dry machine with an isothermalized hot cylinder, or one in which the hot cylinder was small enough to behave isothermally, would therefore show a much lower minimum operating temperature than would the adiabatic expansion cylinder machines considered here.

If it is not isothermalized, the smallest (16-mm-diameter) engine will not operate as a dry machine at all, even at the highest temperature considered (327°C). However, it will operate as a wet machine even at the much lower temperature of 90°C.

The 50-mm-diameter engine should work as a wet machine at low temperature, and it should also work as a dry machine—although with little power—if the temperature is raised sufficiently.

These results are best considered as no more than broad observations, since the numerical details are not at all reliable. The neglect of hysteresis losses in wet machines is an especially suspect assumption. Nevertheless, it appears that the calculations are at least consistent with experience: small machines need evaporation (wet operation) to run at all, and dry machines need a much higher temperature than do wet ones if they are to operate successfully. Note, however, that if the dry machines had their hot cylinders isothermalized, so that expansion-space hysteresis losses were reduced or eliminated, they too might run

Table 8.2. Losses and Output for Wet and Dry Machines (120° Phase Angle).

	DRY						WET		
	600 K			363 K			363 K		
Cylinder diameter, mm	150	50	16	150	50	16	150	50	16
Frequency, Hz†	0.5	0.85	1.2	0.5	0.85	1.2	0.5	0.85	1.2
Tuning-line diameter, mm*	60	25	5	60	25	5	60	25	5
Ideal output, W	44	2.8	0.13	12.6	0.8	0.04	32	2.0	0.09
Adiabatic loss, W‡	3	0.1	0.01	4.7	0.3	0.01			
Hysteresis loss, W‡	14	2.0	0.24	7.0	1.0	0.12			
Flow losses, W	10	0.4	0.06	10	0.4	0.06	10	0.4	0.06
Net output, W	+17	+0.3	−0.18	−9.1	−0.9	−0.15	+22	+1.6	+0.03

*Chosen to correspond very roughly with some of the machines described in Chapters 4 and 9 for which measured performance figures are available.
†Generally, the smaller the machine, the higher the operating frequency (which is set primarily by the displacer length) can be.
‡No hysteresis or adiabatic loss is attributed to the wet machines, on the grounds that evaporation, aided by extra heat-transfer surface if necessary, will help to keep the hot space isothermal. This assumption is not known to be valid.

at lower temperatures; but for the reasons given earlier, the dry machine is not easily isothermalized in practice.

Another major difference between wet and dry machines, which gives some indication of their overall advantages and disadvantages, may be found by considering their heat requirements and hence efficiencies.

HEAT INPUT

We have already estimated the cyclic heat input required by each of the two Fluidyne systems; we must now take into account the major parasitic heat losses. Since we have seen that the dry machines will not run at low temperatures without isothermalization, the comparison is limited to two cases: high-temperature dry machines and low-temperature wet engines (Table 8.3). The figures are calculated from the equations given earlier, assuming that the displaced liquid is water and that the working fluid is air.

Note that for the dry machines, most of the heat input is needed for the parasitic losses. For the wet machine, on the other hand, the cyclic heat input—the heat needed to evaporate the water from the hot cylinder—dominates, even though the use of a liquid piston with no float increases the shuttle loss by a large factor.

Once again, the figures should be taken as no more than a broad guide to the relative magnitude of the various heat losses in wet and dry

Table 8.3. Major Component of Heat Input.

	DRY, 600 K		WET, 363 K		
Cylinder diameter, mm	150	50	150	50	16
Frequency, Hz	0.5	0.85	0.5	0.85	1.2
Cyclic input, W	88	5.6	1850	120	5.4
Insulation heat leak, W*	82	27	17	6	1.8
Shuttle loss, W†	22	2.4	130	16	0.8
Reheat loss, W‡	36	2.2			
Total heat input	228	37.2	1997	142	8.0

*Insulation one-cylinder-radius thick with a conductivity of 0.1 W/(m)(K) assumed, and a heat loss from a region two diameters long.
†Dry machines have a float, with an annular gap of 3 mm to the cylinder wall, to inhibit evaporation.
‡A regenerator ineffectiveness of 0.15 is assumed for the dry machine. No regeneration is incorporated in the wet machine, i.e., all the heat used to evaporate water (except for that part converted to mechanical work) is rejected at the cold end.

machines. With their aid, however, we can estimate the overall efficiency that might be achieved by machines of various sizes and types.

OVERALL EFFICIENCY

Using the power outputs and inputs from Tables 8.2 and 8.3, we may calculate the overall efficiency of the high-temperature dry machines and the low-temperature wet machines. The results are summarized in Table 8.4. Figures calculated earlier have been rounded before being used in Table 8.4.

EXPERIMENTAL MEASUREMENTS

Because only the major losses have been taken into account, the efficiency and output figures shown in Table 8.4 are expected to be optimistic. They are perhaps best regarded as representing an upper limit that may be approached but not substantially exceeded by machines of these sizes and types, although the use of super-insulation, unusually wide tuning lines, isothermalization and higher operating temperatures could lead to machines of still higher performance. To that extent, they are quite consistent with observation; as Table 8.5 shows, the best published results for the efficiency of similar machines are typically about one-third to two-thirds of the figures estimated above.

We can draw the following conclusions from Table 8.5. For a certain range of cylinder sizes, more than 50 mm (2 in) in diameter, the dry machine will be much more efficient than the wet system but will give lower output power. For still smaller sizes, the losses will be so high relative to gross ouptut that a dry machine may not run at all with any reasonable heater temperature; therefore, wet operation is essential for

Table 8.4. Overall Efficiency of Wet and Dry
Machines.

	DRY, 600 K		WET, 363 K		
Cylinder diameter, mm	150	50	150	50	16
Net output, W	17	0.3	22	1.6	0.03
Heat input, W	230	37	2000	140	8
Efficiency, percent	7.0	0.8	1.1	1.1	0.4

Table 8.5. Estimated Performance Limits and Measured Performance of Wet and Dry Machines.

DIAMETER, mm	TYPE	NET POWER, W		EFFICIENCY, PERCENT		REFERENCE
		ESTIMATED UPPER LIMIT*	MEASURED†	ESTIMATED UPPER LIMIT*	MEASURED†	
150	Dry	17	14	7.0	5	West and Pandey (1981)
150	Wet	22		1.1		No experience
50	Dry	0.3	>0	0.8	>0	Goldberg and Rallis (1979)‡
50	Wet	1.6	1.7	1.1	0.35	West, 1971
50	Wet	1.6	0.2	1.1	0.5	Reader et al. (1981)
16	Dry	0	0	0	0	Ryden, unpublished§
16	Wet	0.03	0.02	0.4	0.15	Mosby (1978)
16	Wet	0.03	0.03	0.4		West (1970a)

*These estimates are calculated from Tables 8.3 and 8.4 under the assumptions made in this chapter. Hot end temperature is 600 K (dry) or 363 K (wet).
†The actual machines with these performance figures had approximately the cylinder diameters shown, but may have differed in other respects from the simplified models analysed in this chapter.
‡Results obtained with a solid-displacer, liquid-output-piston engine.
§In an unpublished experiment at Harwell, Ryden found that a machine with a 15- to 20-mm-diameter cylinder would not operate when the water in the displacer was replaced with a low-vapor-pressure liquid.

small engines, but efficiency will be low. This conclusion may be modified if the hot cylinder of the dry machine is isothermalized, when the lower limit to the size of a self-sustaining dry engine will be reduced.

For larger machines, the thermal losses associated with evaporation in wet systems dominate other heat losses. For good efficiency, a large machine operating on a dry cycle is essential; however, the power density or specific power is still lower than could be obtained with wet operation.

With this information in mind, we may speculate on the most likely application areas for the two types of machine. Where the cost of heat is high and a large continuous output is needed, dry machines are the most likely choice. Examples are irrigation and farm drainage pumping using fossil fuels, concentrating solar collectors or limited resources of biomass, especially in poor societies where the cash values of the crop may be low compared with the cost of purchased energy.

On the other hand, if the cost of heat is low and it is available only at low temperatures, as it may be if waste heat or even flat-plate solar collectors are to be used, then a wet machine, or a dry machine with an isothermalized hot cylinder and a filling of low-vapor-pressure liquid, may be used.

If a small size and good throughput are needed in circumstances that make the cost of heat negligible (such as bilge pumping), then a wet machine, with its higher power density, is likely to be preferred to a dry system.

These conclusions depend upon the Walker and Agbi analysis which, for simplicity, assumes that the proportion of water vapor and air does not change during the cycle and is the same throughout the engine. It may be possible to avoid this limitation in the Fluidyne, where the flow is generally not turbulent and little mixing of the gas takes place: if so, it would be possible to raise both the power and the efficiency considerably above the performance of the wet Fluidynes discussed here.

Chapter 9
Past Research Work and Some Present Research Needs

INITIAL HARWELL WORK

The Fluidyne principle was first devised at Harwell in 1969 by C. D. West but remained an idea on paper until 1970, when effort became available, with the arrival at Harwell of a summer student (J. M. McDonald), to carry out some experimental work. During the course of that summer, the principle was experimentally verified and several different machines were built. Little further work was done or progress made until 1974, when the 1970 results were declassified and published.

The 1970 Harwell results are described in two reports (West, 1970a; and West, 1971) that were declassified in 1975 and 1974, respectively, at which time several other research groups and individuals began to work on the Fluidyne and to investigate various aspects of its behavior. These two reports describe the basic principles of the Fluidyne and give experimental confirmation of both rocking beam feedback and liquid feedback. (One form of liquid feedback, shown in Figure 2.6, was proposed by Cooke-Yarborough in 1970 and was the first liquid feedback system to be tried; it worked successfully and has been adopted by other workers in the field as the most flexible feedback method and the least sensitive to mistuning.) One of these reports (West, 1971) includes details of a liquid feedback engine that pumped 100 U.S. gal/h through a head of over 5 ft. Evaporation was not suppressed in these machines, and the efficiency was relatively low: a maximum efficiency of 0.35 percent was measured for the 100 gal/h machine. Some of the theory presented in those reports is wrong, largely because the effects of oscillating flow on viscous losses and on heat conduction down the liquid column were not known to the writer at that time. In addition, neither the transient heat-transfer loss nor the full significance of evaporation in increasing both the power output and heat loss were known. It is perhaps surprising that the earliest machines worked at all in face of such ignorance, but the fact is that the small machines with evaporation proved to be very easy to operate and to be remarkably insensitive to mistuning.

HARWELL AND METAL BOX PROGRAM

In 1974, the Metal Box companies became interested in the Fluidyne as a marketable product for irrigation pumping in developing countries. They sponsored a small research program at Harwell for the further development of the Fluidyne, and assigned an engineer (Ram Pandey) from the Metal Box Company of India to work with Harwell staff. In early 1977 the Harwell program was completed, having produced results that showed the potential of these machines for irrigation pumping. The program was transferred to Calcutta, the headquarters of Metal Box India, and continued at a low level (with perhaps only two or three technical staff members) over the next 3 years. By the end of 1980, the company had become convinced that adequate penetration of the market area they were primarily aiming at—India—required a pumping head that could not be achieved by their present Fluidyne design without adding complications to the basic machine that would unacceptably increase its costs, and the company began seeking ways to continue development work on machines with a greater pumping head. Other markets could be satisfied with the capability (10-ft head) of present machines, and other companies have shown an interest in their manufacture. Metal Box has recently published some operating results from their in-house experimental development program. One machine with a concentric cylinder layout gives a pumping rate of 2500 gal/h at a head of 10 ft with an efficiency of 7 percent (Pandey, 1981a). With a gas burner, a pumping rate of 4000 U.S. gal/h at a head of 12 ft has been achieved (Pandey, 1981b); the overall efficiency in this case, including burner efficiency, was $3\frac{1}{2}$ percent. More detailed results are available from the earlier program carried out at Harwell on behalf of Metal Box (West and Pandey, 1981).

Prior to 1981, however, no results from the Harwell–Metal Box program were published, so that other researchers in the 1970s worked independently of these developments, with access only to the 1970 Harwell results.

During the course of the Metal-Box-sponsored work at Harwell, there were very significant advances made in the theoretical understanding of various effects in the engine, particularly by Ryden, whose work (not published) included an exploration of the behavior of oscillating liquid columns and the identification of the transient heat transfer loss—although this effect had, unknown to the Harwell group, been reported earlier (Breckenridge, Heuchling, and Moore, 1971) in a limited distribution U.S. government document released to the general public in 1975. The Breckenridge, Heuchling, and Moore results referred to losses in a gas spring, and the concept was later extended by Lee, Smith, and Faulkner (1980) to include Stirling machine power losses. Ryden also proved the importance of evaporation to the operation of most previous machines. Even prior to this proof, it was already known that evaporation might limit overall efficiency to less than 1 percent in air-filled machines; and since the target performance of the program called for a 3 percent efficiency, effort was being concentrated on dry machines in which evaporation was suppressed. This efficiency target was exceeded and the target pumping head of 10 ft was also achieved, but the laboratory machine's throughput of 460 U.S. gal/h was less than half the design figure because the observed phase angle between piston motions was much larger than expected. Subsequent theoretical work (West, 1978) established that the reason was probably a fairly large loss almost in phase with the tuning column motion, but no experimental data were available to identify the mechanism for such a loss. Two factors may be major contributors to the excess loss. First, the existing formulas might be seriously underestimating the flow losses (see Chapter 5) and these could therefore be a major contributor to the excess losses. Second, it now appears that the transient heat-transfer loss in nearly adiabatic spaces is 5 to 10 times larger than may be predicted from a simple conductivity model (Lee, Smith, and Faulkner, 1980; Wood, 1980); the earlier report of an enhancement factor (Breckenridge, Heuchling, and Moore, 1971) was not declassified at the time the Metal Box machine was designed and tested; and in any case, the report dealt with a situation where the enhancement factor was only one-quarter to one-half this value. The larger enhancement factor has very important consequences, especially for low-speed engines (see

Chapter 7), and probably provides a link in the chain of explanation for the high phase angle of the Harwell–Metal Box laboratory prototype engine.

After the results obtained at Harwell, Metal Box continued with the development of the Fluidyne to higher efficiency and much higher throughput. However, there is no known way in which a simple, single-stage, atmospheric-pressure Fluidyne can achieve the pumping head (20 ft) that is sought by the company for their chosen market. All known ways of increasing the pumping head to 20 ft or more [such as pressurization of the working fluid (West, Geisow, and Pandey, 1977) or staging the pumps (Figure 9.1)] add considerably to the complication

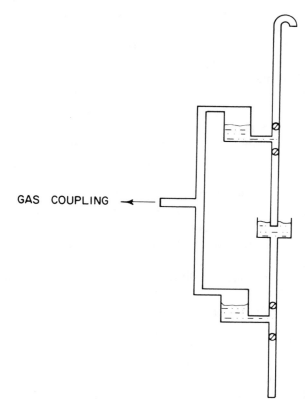

GAS COUPLING

Figure 9.1 Driving a two-stage pump from a single displacer in order to increase pumping head.

and hence the cost of the machine, besides detracting from the simplicity and ease of maintenance that are part of the attraction of the Fluidyne concept.

UNIVERSITY OF THE WITWATERSRAND

Staff and students in the Department of Mechanical Engineering at the University of Witwatersrand have also investigated the Fluidyne system. Their early results were obtained from an engine whose design was based roughly on the 100 gal/h Harwell engine. Their results show (Bell and Goldberg, 1976; Goldberg, Rallis, Bell, and Urieli, 1977) that although their engine would operate as either a wet or dry Fluidyne (i.e., with or without evaporation), useful power could be extracted only when evaporation was permitted. This work was continued and led to a very complete description of various phenomena observed with wet machines, and some assessment of the differences between wet and dry operation (Goldberg, 1979). It was noted that the choice of dimensions, i.e., the tuning, was much more critical for dry than for wet operation. On the other hand, the orientation of the junction between the tuning column and the displacer (pointing toward or away from the hot end) was found to be crucial in the case of this wet engine, although this is known to be less important in dry machines constructed at Harwell. It is not clear whether this reflects a fundamental difference between wet and dry machines, or whether it is a size effect: Lewis (1978) states that momentum effects will become relatively less important as the junction between the output column and the displacer becomes larger. Goldberg, in his thesis, concludes that a Fluidyne of the size he studied was not a viable energy-conversion device: wet operation led to large enthalpy leaks and low efficiency, while dry operation gave too little power output to be useful. He emphasizes that these conclusions were drawn from experience with a particular size engine and might not be valid for larger machines. (As the Metal Box–Harwell results show, scale-up does indeed permit much better performance.)

In view of the specific power limitations of the Fluidyne that they identified, the Witwatersrand group began development of a liquid piston version of a beta-configuration free-displacer engine with a liquid column replacing the solid power piston (Goldberg, 1979; Goldberg and Rallis, 1979). The use of a solid displacer piston effectively eliminates the enthalpy flux leakage between the hot and cold spaces and also permits improved efficiency by the accommodation of higher hot-space temperatures. The displacer was driven by the pressure differential between the working and bounce spaces, the latter, in the initial configuration, being isolated from the liquid column. The configuration was proposed by Rallis and based upon a generically similar engine conceived by Martini (1977). It was developed independently of the similar solid displacer Fluidyne proposed earlier at Harwell (West, Cooke-Yarborough, and Geisow, 1970).

Using a coupled second-order computer simulation program (involving inertial and viscous liquid column dynamics) as a design tool, a series of experimental prototypes were built and tested by the Witwatersrand group. The final prototype developed included a means of coupling the working and bounce spaces which resulted in a three- to fourfold performance improvement over the initial prototype. All the prototypes exhibited a performance improvement over a comparably sized machine with all liquid pistons. A large number of variants of free-displacer liquid piston engine, including some multicylinder engines and different work-extraction methods, were also proposed (Goldberg, 1979).

In 1979 the liquid piston engine program at the University of Witwatersrand was continued with the development of a liquid piston gamma version of a back-to-back free-displacer engine invented by Beale. This gamma configuration was proposed by Rallis and a successful laboratory prototype was built and tested (Leigh, Lohrman, and Swerdlow, 1979; Goldberg, 1980). In addition, work was continued on the all liquid piston machine, an aluminum float being inserted in the hot-space column. This improved the efficiency but reduced the power output.

ROYAL NAVAL ENGINEERING COLLEGE

The Stirling Engine Research Facility (SERF) of the Royal Naval Engineering College has recently carried out a number of experimental and theoretical investigations of the Fluidyne system. Most of their experimental results were obtained with the wet Fluidyne system (Reader and Lewis, 1979a) and showed the same kind of efficiency (1 percent) as other evaporative Fluidynes. They noted that the desinent threshold for oscillations could be as low as 20 K temperature difference between the hot and cold cylinders (T_e = 100°C, T_c = 80°C), although self-starting did not occur until the temperature difference was 70 to 80 K. They also observed that in their machines the oscillations, once self-started, did not increase gradually but rather started abruptly with a finite amplitude. Both these effects are characteristic of a system exhibiting, in mathematical terms, hard self-excitation. Similar effects were noted at Harwell, although one small machine (which may have been small enough for the cylinders to behave isothermally) was found to have a threshold of only 25 K even at low temperatures (T_e = 25°C, T_c = 0°C) where evaporative effects are presumably small.

Another paper (Reader and Lewis, 1979b) identifies three Fluidyne operating regimes revealed in their experiments: these are described in Table 9.1.

Another observation made by the SERF group is that for evaporative operation on their particular machine, there were large differences, both in phase and amplitude, between the pressure variations in the hot and cold cylinder (Reader, 1979). Neither pressure variation was sinusoidal, a fact they attributed to the presence of water vapor. The liquid-column movements were also markedly nonsinusoidal in their apparatus.

The SERF work has also included the development of a mathematical model, with load and viscous terms, of the stable oscillations of the Fluidyne, and its implementation on a computer.

These, and other projects, are described in some unclassified, but hard to acquire, reports from the Royal Naval Engineering College (Lewis, 1978; Lloyd and Hensman, 1979; Singleton, 1979; Thwaites, 1979; Gill, 1980). Lewis's computer program, as extended and corrected by Gill, predicts correctly some of the observed behavior of their

Table 9.1. Operating Regimes for Atmospheric Pressure Fluidynes.

TYPE OF MACHINE	HOT-SPACE TEMPERATURE, K	COLD-SPACE TEMPERATURE, K	COMMENTS
Wet	370	300	Air cycle is dominant
Wet	470	320	Liquid-vapor cycle and closed regenerative (air) cycle are simultaneously evident
Dry	500	350	Closed regenerative cycle using dry air is dominant

experimental Fluidyne. However, the computations do involve an empirically determined "feedback factor," the ratio of liquid flow rate in the displacer column and output column. The theory does not include any derivation of this crucial factor, but there is a useful discussion of the relative importance of pressure terms and convective acceleration, i.e., pressure feedback effects versus jet-stream feedback effects. Hensman's report (1980) also contains a derivation of the dynamic equations including viscous and load terms, based on the work of Stammers (Stammers, 1979) and some experimental results. Once again some broad trends are correctly predicted. However, many of the experimental results (especially efficiency) could not be used directly for comparison with theory or with other results, because the pumping system used would not operate continuously in a stable manner.

MATHEMATICAL ANALYSES

The original analysis of the most commonly used liquid feedback system (Elrod, 1974) aimed at deducing the stability criteria of an ideal system: that is, it established the temperature difference required for infinitesimal oscillations to begin growing. The linearized mathematical model employed was recognized as being inapplicable to large oscillations and no damping or load terms were included. Nevertheless, this analysis provided insights that were invaluable in the development of liquid feedback systems. Elrod's linearized analysis was subsequently programmed onto an analog computer (Gosling and Boast, 1976), which proved to be a useful and powerful way of studying some of the properties of these equations. Geisow's published analysis (1976) used the same mathematical model, but in unpublished work he was able to extend it to include the effects of some damping terms. Recently, a paper by Stammers (1979) reported an analysis that admitted viscous losses and a velocity-dependent load.

Stammers' results are very important in understanding some of the effects observed in real machines, especially dry engines. An important parameter in determining the threshold for oscillations to begin, and in determining the behavior of a loaded engine, is the difference in length of the cold-displacer column and the hot-displacer column, both being measured from the junction between the tuning column and the displacer. In Elrod's analysis, the temperature difference for oscillations to begin is minimized by minimizing the difference in these lengths. Experience with the Metal Box–Harwell prototype machine was that better performance is obtained by maximizing this difference, at least within the range of difference that was experimentally feasible. Stammers' analysis both predicts and explains this effect, and also shows that exact tuning of the lengths becomes less important as the engine is more heavily loaded. This latter is an important prediction because the tuning line is a bulky, expensive, and lossy component of the system, and if it can be shortened without much loss of performance, there is much to be gained. Few data are available in this area, besides those of Stammers, but some measurements of the effect of mistuning on certain performance characteristics of a wet Fluidyne are reported by Findlay and Hook (1977). The report of a computer study by Drzewiecki (1979) included some discussion of the effect of tuning-line length on a lossless machine without evaporation.

OTHER RESULTS

Mosby (1978) reports results obtained with a pump constructed according to the design shown in Figure 4.6. He was able with this small machine to pump 5.9 U.S. gal/h at a low head (10 cm) and achieved an overall efficiency of 0.15 percent. Bell (1979*a, b*) built and tested a wet Fluidyne with the eventual intent of using a solar-powered machine for irrigation water pumping: he achieved a maximum pumping rate of about 30 U.S. gal/h with a head of 4 ft and a peak efficiency of 0.18 percent.

There has been very little work on multicylinder configurations, the only known published experimental results being obtained by researchers at the Chicago Bridge and Iron Company (Cutler and Hanke, 1979). They tested two machines, in both of which there was considerable evaporation. They note that analysis of the system is greatly complicated by the need to take into account the boiling effect of the pistons. Table 9.2 summarizes some of the published performance data on Fluidyne pumps.

RESEARCH NEEDS

In this section, we discuss briefly some of the research needs and opportunities that lie specifically in the field of liquid piston Stirling engines. Research needs that relate to Stirling systems in general (such as a bet-

Table 9.2. Some Published Performance Figures for
Fluidyne Pumps.

| | PERFORMANCE | | |
REFERENCE	FLOW RATE, U.S. gal/h*	HEAD, ft*	EFFICIENCY, PERCENT*
West (1970)	3	3.3	
West (1971)	100	5.3	0.35
Goldberg et al. (1977)	9.5	2.0	0.12
Goldberg et al. (1977)	11.5	3.0	0.08
Mosby (1978)	5.9	1.0	0.15
Reader (1979)	. . .	. . .	.03
Bell (1979)	30	3.6	0.18
Reader et al. (1981)	2.0	3.3	0.52
West and Pandey (1981)	460	10.0	4.7
Pandey (1981a)	2500	10.0	7.0

*Figures quoted are measured performance characteristics but were not
necessarily achieved simultaneously.

ter knowledge of the transient heat-transfer losses in both nearly adiabatic and nearly isothermal spaces) are implicit throughout this book.
Crankshaft and, to a lesser extent, free piston engines have been established longer and have received more development attention than have
liquid piston systems; consequently, the remaining research projects for
solid piston engines tend to be more difficult and expensive. There are
many important but simple questions to be asked about Fluidyne systems, and the research necessary to answer many of these questions is
within the range of even small projects, including university student
projects. There are few enough fields in which small, short-term projects
can break new ground, but this is one of them. In this respect, some
suggestions for simple but useful experiments were given in Chapter 4,
and some more suggestions are given here.

Almost all the experimental data have been obtained on liquid feedback machines (an exception being the free-displacer engine results
from the University of Witwatersrand). Nevertheless, as we have seen,
there are some potential advantages to the rocking beam system, including a wider choice of operating frequency and a greater freedom in
selecting the liquid that forms the hot piston. A thoughtful (but not
excessively numerical) analysis of the rocking beam machine would be
a valuable contribution to the field. Some experimental measurements

on a well-designed rocking beam pump of reasonable capacity—say 100 gal/h or more—would be even more valuable.

An engine with a stationary tuning line and the displacer rocked externally by a variable speed and amplitude driver would be a good tool for investigating the effects of varying frequency and displacer amplitude.

For liquid feedback engines, more work is badly needed on low-vapor-pressure liquids that are immiscible with water for use in the hot cylinder, so that the cylinder can be isothermalized and transient heat-transfer losses reduced.

The importance of transient heat-transfer losses in these low-speed machines may mean that the optimum compression ratio in practice is lower than the theoretically ideal value of around 2. A pencil-and-paper investigation of the effect of reducing compression ratio by increasing the dead space would be a very useful exercise. Such an analysis has been carried out for the diaphragm free piston engine with the intent of optimizing regenerator efficiency and diaphragm stress, but not taking into account transient heat-transfer losses (West, 1970*b*).

More data are needed on losses, including the kinetic losses, in oscillating flow. Although experimental work has been carried out [e.g., Biery (1969)] only a few results are available (Chan and Baird, 1974) for the combination of tube diameter and amplitude appropriate to a large Fluidyne, and these do not include the effects of kinetic losses. One report based on work with narrow tubes indicates that change of cross section may lead to very severe flow losses (Richardson, 1963), and this is obviously important in Fluidyne design, especially if the effect applies equally to large bore tubes. Little more than a collection of large, transparent U tubes, a stop watch, and a scale (or better still, a movie camera) is needed to begin work on this problem—if researchers choose their experiments carefully.

There is an advantage in the form of reduced reheat losses and more isothermal behavior in the cold cylinder to using narrower regenerator tubes or screens and isothermalizers. The limit is set by increased flow losses and, probably more importantly, by blockage of the narrow spaces with water from the liquid piston or from condensation. Some very simple experiments on blockage of different-sized tubes, annuli, and screens could provide new and very useful data for the Fluidyne designer.

Most of the early published results on Fluidynes were obtained with wet machines. This led many people to believe that the efficiency of the Fluidyne is limited to a small fraction of 1 percent, but with the publication of the Harwell–Metal Box results we see that there is no such limitation. Furthermore, from a strictly research point of view, permitting evaporation greatly complicates the theoretical analysis of engine behavior. Experimental data from a high-performance Fluidyne, operating without evaporation, are needed if the existing theory is to be tested; and until it is tested, there seems little point in looking for more sophisticated theories.

As indicated in Chapter 8, the thermodynamics of wet machines with a mixed air–water vapor working fluid have hardly been explored at all. The potential advantages of wet machines in certain circumstances (low-temperature operation, small size) are quite important, but the design theory available is even less complete than that for dry machines. Better ways of estimating the effects of evaporation and of optimizing wet machines are badly needed.

There is a great temptation to build extensive computer models. Please do not fall for the temptation. Computer models do not pump water onto the laboratory floor, burn out their heaters, or leak ambient air into their hot cylinders, but neither do they give the final stamp of approval to a new idea; only experiment can do that. We still do not know enough about the simple things to try to learn more from very complex analyses of the Fluidyne.

What aspects of performance, as distinct from understanding, are most in need of improvement? The most obvious application of the Fluidyne is still that of a simple, low-cost water pump. For many such uses, the throughput and efficiency of the Metal Box designs are adequate, but the geographical range of application could be greatly extended if a simple and low-cost way of increasing the pumping head could be devised. More work is needed on pressurized machines and alternative methods for increasing the pumping head.

The tuning line is a large and costly component in large-scale Fluidynes. An investigation of the effects of using a tuning column much shorter than the resonant length, and perhaps much narrower than the present designs, or of alternative tuning methods, could lead to substantially reduced bulk and manufacturing costs for Fluidyne water pumps.

Appendix
Gas-Coupled Pumps

Consider the gas-coupled pumping system shown in Figure 2.18. Until the pressure (or suction) inside the gas space exceeds the hydrostatic force exerted on the valves by the water in the pumping lines, the valve cannot open and no water can flow. Thus if the pump is arranged symmetrically, the outlet valve will open only if

$$\Delta P \sin \omega t \geq \frac{H}{2} \rho g \qquad (A.1)$$

where ΔP = pressure amplitude in working gas
ω = angular frequency of operation
H = pumping head (assumed to be symmetrically disposed above and below the gas-coupling connection)
ρ = density of liquid being pumped (usually water)
g = acceleration due to gravity

The valve will open when

$$\omega t \geq \sin^{-1} \frac{H \rho g}{2 \Delta P} \qquad (A.2)$$

At the moment of equality, we set $\omega t = \theta'$ and $t = t'$. Once the valve has opened, the pressure across the water column, $\Delta P \sin \omega t - (H/2) \rho g$, begins to accelerate the water.

$$\text{Acceleration} = \ddot{x} = \frac{\Delta P \sin \omega t - (H/2) (\rho g)}{(H/2) \rho} \bigg|_{\omega t \geq \theta'} \qquad (A.3)$$

133

The velocity of the water, $\dot{x}$, is therefore

$$\dot{x}(t)\mid_{t \, > \, t'} \; = \; \int_{t'}^{t} \ddot{x} \, dt \; = \; \int_{t'}^{t} \left(\frac{2 \, \Delta P}{gH\rho} \sin \omega t - g \right) dt \tag{A.4}$$

$$= \left[\frac{-2 \, \Delta P \cos \omega t}{\omega H \rho} - gt \right]_{t'}^{t}$$

Substituting for t' from Equation (A.2),

$$\dot{x}(t) \; = \; \frac{2 \, \Delta P}{\omega H \rho} \left[\cos \left(\sin^{-1} \frac{H\rho g}{2 \, \Delta P} \right) - \cos \omega t \right]$$
$$- g \left[t - \frac{\sin^{-1} \left(H\rho g / 2 \, \Delta P \right)}{\omega} \right] \tag{A.5}$$

The maximum velocity occurs when the acceleration is just reversing, i.e., when $\omega t = \pi - \theta'$ and the internal pressure has once again fallen to the hydrostatic pressure. However, the valve remains open due to the momentum that the water column has already acquired. We may suppose that the valve does not close until the velocity of the water has again fallen to zero. This happens at a time t'' and phase angle $\omega t''$, which we call θ'', where

$$\frac{2 \, \Delta P}{\omega H \rho} \left[\cos \left(\sin^{-1} \frac{H\rho g}{2 \, \Delta P} \right) - \cos \omega t \right] \tag{A.6}$$
$$= g \left[t - \frac{1}{\omega} \left(\sin^{-1} \frac{H\rho g}{2 \, \Delta P} \right) \right]$$

Therefore

$$\theta'' \; = \; \omega t'' \; = \; \cos^{-1} \left[\cos \left(\sin^{-1} \frac{H\rho g}{2 \, \Delta P} \right) \right. \tag{A.7}$$
$$\left. - \frac{H\rho g}{2 \, \Delta P} \left(\theta'' - \sin^{-1} \frac{H\rho g}{2 \, \Delta P} \right) \right]$$

This equation cannot be solved explicitly for θ'', but it can be easily solved graphically or with the aid of a programmable calculator.

Now that we know the velocity of the water and the time for which the valve is open, it is easy to calculate the volume of water in each stroke, V_{pumped}:

$$V_{pumped} = \int_{t'}^{t} A_p \dot{x}(t) \, dt \qquad (A.8)$$

where A_p is the cross-sectional area of the inlet and outlet pipes on the pump. Substituting for $\dot{x}(t)$ from Equation (A.5) yields a value for the pumped volume:

$$V_{pumped} = A_p \frac{g}{\omega^2} \left\{ \frac{2 \, \Delta P}{H \rho g} \left[(\theta'' - \theta') \cos \theta' - (\sin \theta'' - \sin \theta') \right] - \frac{(\theta'' - \theta')^2}{2} \right\} \qquad (A.9)$$

$$= A_p \frac{g}{\omega^2} \left\{ \frac{(\theta'' - \theta')}{\tan \theta'} + 1 - \frac{\sin \theta''}{\sin \theta'} - \frac{(\theta'' - \theta')^2}{2} \right\} \qquad (A.10)$$

Note that in this equation, ΔP appears as though it were a constant of the system. This is not the case, for the movement of water into and out of the pump arm will change the volume and hence the pressure of the working fluid. Generally, as the pumped volume per stroke increases, the pressure amplitude in the machine decreases. For this reason, the pumped volume does not continue increasing indefinitely as the area of the pumping line is increased, although this is, at first sight, the behavior that seems to be predicted by Equation (A.10). For a numerical estimate of this effect, it is possible to construct a load line that relates the pumped volume and the pressure amplitude, but that task is beyond the scope of this appendix.

Example

Even with the assumption that ΔP is constant, the flow rate is very dependent on the pressure amplitude, as we may see by calculating the results for a particular example. Consider a machine pumping water

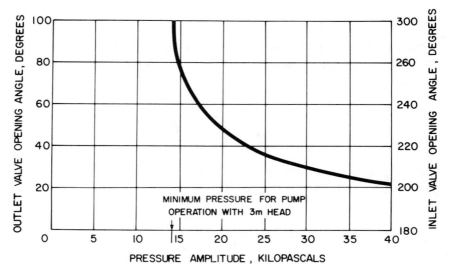

Figure A.1. Valve opening angle as a function of pressure amplitude.

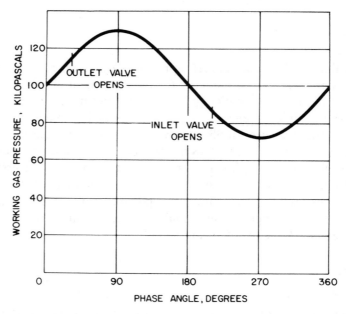

Figure A.2. Working gas pressure and valve opening times for 3m pumping head and 30kPa pressure amplitude.

through a head of 3 m (approximately 10 ft). From Equation (A.2), the phase angle (relative to the pressure variation in the working fluid) at which the pump valves first open is $\theta' = \sin^{-1}[(3 \times 10^3 \times 9.81)/(2 \times \Delta P)] = \sin^{-1}[1.47 \times 10^4/\Delta P]$; this relation is illustrated in Figure A.1. Only if the pressure amplitude exceeds about 15 kPa—which corresponds to a pressure ratio of approximately $(100 + 15)/(100 - 15) = 1.35$—will the pump work at all. For a pressure amplitude of 30 kPa, the valves would open about 30° after the pressure has passed through its average value; i.e., about 60° before the pressure reaches its maximum (outlet valve) or minimum (inlet valve): this situation is illustrated in Figure A.2.

We must next calculate the phase angle at which the valves close, by finding the value of ωt at which the velocity has fallen to zero [see Equation (A.7)]. Figure A.3 illustrates this by showing the flow rate as a function of phase angle for a pumping head of 3 m, pressure amplitude of 30 kPa and a frequency of 0.65 Hz, or 4.08 r/s. In this particular case, the flow velocity falls to zero again at a phase angle of 220°,

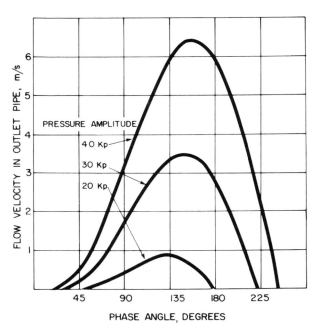

Figure A.3. Flow velocity in the outlet pipe as a function of phase angle.

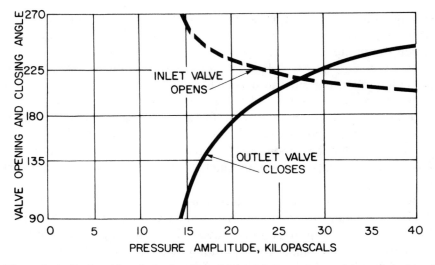

Figure A.4. Outlet valve closing angle and inlet valve opening angle as a function of pressure amplitude.

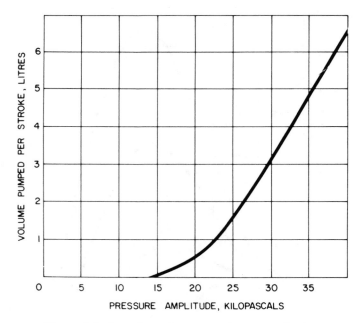

Figure A.5. Pumping rate versus pressure amplitude.

and the valve then closes. The figure also shows the velocity-time profile for some other values of the pressure amplitude.

Figure A.4 gives the valve-closing angle as a function of pressure amplitude, again assuming a total head of 3 m. At high values of the pressure amplitude, an interesting thing happens: the outlet valve does not close until after the inlet valve has opened, and vice versa. As indicated in Figure A.4, for a pumping head of 3 m this happens when the pressure amplitude exceeds about 27 kPa. For all higher values of the pressure amplitude, there are times when both valves are open simultaneously. For lower pressure amplitudes, there are times when neither valve is open. This second condition is usually observed in practice, and there have as yet been no published reports of the first case: perhaps no one has been able to make a Fluidyne machine operate with this combination of pumping head and compression ratio, or perhaps no one has looked for the effect.

Now that we have the valve opening and closing times, we can calculate the pumping rate. For this, we need to know the pumping line diameter and the frequency of operation. We assume a diameter of 50 mm and a frequency of 0.65 Hz or 4.08 rad/s. Substituting these figures and the appropriate phase angles already derived into Equation (A.10) gives the pumping rate for a 3-m head as a function of pressure amplitude (see Figure A.5). The volume pumped per stroke increases rather rapidly as the pressure amplitude increases: note, however, that a rather large machine would be needed to maintain a pressure amplitude of, say, ± 25 kPa with a volume variation due to the pump of almost 2 L. We have also ignored both the pressure head and the added mass of the water in the pump sidearm.

Bibliography

Allen, P. C., Knight, W. R., Paulson, D. N., and Wheatley, J. C. (1980). Principles of Liquids Working in Heat Engines. *Proc. Nat. Acad. Sci. USA,* vol. 77, no. 1, January.

AMTI (1982). "Analytical Study of Liquid Piston Heat Pump Technology." Private communication, from Argonne National Laboratory, January.

Bell, G. C. (1979*a*). "Passive Solar Water Pump." Independent project, Arch. 589, University of New Mexico.

Bell, G. C. (1979*b*). "Solar Powered Liquid Piston Stirling Cycle Irrigation Pump." SAN-1894/1, April.

Bell, A. J. and Goldberg, L. F. (1976). "The Fluidyne Engine." Final-year laboratory project, University of the Witwatersrand.

Biery, J. C. (1969). "The Oscillating Manometer: Review of Experimental, Theoretical and Computational Results." *J. AIChE*, vol. 15, no. 4, pp. 631–634, July.

Breckenridge, R. W. Jr., Heuchling, T. P., and Moore, R. W. Jr. (1971). Rotary Reciprocating Refrigeration System Studies, pt. 1, *Analysis.* Arthur D. Little Inc., Tech. Rept AFFDL-TR-71-115, September.

Chan, K. W. and Baird, M. H. I. (1974). "Wall Friction in Oscillating Liquid Columns." *Chem. Eng. Sci.,* vol. 29, pp. 2091–2099.

Cooke-Yarborough, E. H. (1974). "Simplified Expressions for the Power Output of a Lossless Stirling Engine." AERE-M 2437, March.

Cooke-Yarborough, E. H. (1975). "Stirling Cycle Thermal Devices." US Patent 4,077,216 (filing priority date August).

Crandall, I. B. (1927). *Theory of Vibrating Systems and Sound.* Van Nostrand, Princeton, N.J.

Crane Company (1957). "Flow of Fluids through Valves, Fittings and Pipe." Crane Tech. Pap. 410, Engineering Division, Crane Company, Chicago.

Cutler, D., and Hanke, C. (1979). "Test Report of Two Fluid Piston Heat Engines." Chicago Bridge and Iron Co., CBI Res. Contract R-0268.

Dros, A. A. (1965). "An Industrial Gas Refrigerating Machine with Hydraulic Piston Drive." *Philips Tech. Rev.,* vol. 26, no. 10.

Drzewiecki, T. M. (1979). "An Initial Model for the Finite Displacement Response Characteristics of a Fluidyne Pump." HDL-TR-1868, February.

Dunn, P. D., Rice, G., and Thring, R. H. (1975). "Hydraulic and Rotary Drive Stirling Engines with Fluidized Bed Combustion/Heat Pipe System." Pap. 759141. *Proc. 10th IECEC, Newark*, August.

Elrod, H. G. (1974). "The Fluidyne Heat Engine: How to Build One—How It Works." ONR London Rept. R-14-74 (NTIS No. AD/A-006-367), December.

Ewing, Sir J. A. (1926). *The Steam Engine and Other Heat Engines,* 4th ed. Cambridge University Press, New York.

Findlay, R., and Hook, P. (1977). "A Study of a Fluidyne Heat Engine." Thesis, Department of Mechanical Engineering, McGill University, Montreal, Canada. April.

Geisow, A. D. (1976). "The Onset of Oscillations in a Lossless Fluidyne." AERE-M 2840, October.

Gerstmann, J., and Friedman, Y. (1977). Liquid Piston Heat-Actuated Heat Pump and Methods of Operating Same. U.S. Patent 4, 148, 195 (filed December 1977).

Gill, P. F. (1980). "The Mathematical Modelling of a Jet-Stream Fluidyne." RNEC-SERF-F1-80, May.

Goldberg, L. F. (1979). "A Computer Simulation and Experimental Development of Liquid Piston Stirling Cycle Engines." M.Sc. thesis, University of the Witwatersrand, Johannesberg, South Africa, March.

Goldberg, L. F. (1980). "A State Space Analysis of a Symmetrical Compounded Free Piston Stirling Engine." Pap. 809450, *Proc. 15th IECEC, Seattle,* August.

Goldberg, L. F., and Rallis, C. J. (1979). "A Prototype Liquid-Piston Free-Displacer Stirling Engine." Pap. 799239, *Proc. 14th IECEC, Boston,* August.

Goldberg, L. F., Rallis, C. J., Bell, A. J., and Urieli, I. (1977). "Some Experimental Results on Laboratory Model Fluidyne Engines." Pap. 779255, *Proc. 12th IECEC, Washington,* August.

Gongwer, C. A. (1950). "Water Jet Propulsion Systems Without Primary Rotating Machinery." *Symp. Hydraul. Jet Propul., ONR.*

Gongwer, C. A. (1960). "Some Aspects of Underwater Jet Propulsion Systems." J. ARS, pp. 1148–1151, December.

Gosling, M., and Boast, D. (1976). "Analog Simulation of a Fluidyne Engine." B.Sc. Proj. Rept. 382, University of Bath, June.

Hensman, T. W. (1980). "Assessment of a Hard-Coupled Fluidyne Pump." RNEC-SERF-F2-1980, April.

Kays, W. M., and London, A. L. (1964). *Compact Heat Exchangers,* 2d edition. McGraw-Hill, New York.

Kutateladze, S. S. (1963). *Fundamentals of Heat Transfer.* Academic, New York.

Lee, K. P., Smith, J. L., Jr., and Faulkner, H. B. (1980). "Performance Loss Due to Transient Heat Transfer in the Cylinders of Stirling Engines." Pap. 809338, *Proc. 15th IECEC, Seattle,* August.

Leigh, S. W., Lorhman, P. C., and Swerdlow, R. (1979). "The Design, Manufacture, Operation and Preliminary Testing of a Liquid Piston, Free Displacer, Back-to-Back, Gamma Type Stirling Engine." Final-year Lab. Proj., University of the Witwatersrand, Johannesberg, South Africa.

Lewis, P. D. (1978). "Operation of a Jet-Stream Feedback Fluidyne." RNEC-TR-78008, May.

Lloyd, J. (1975). "And Yet It Moves, Again." *New Scientist,* pp. 11, 12, April 3.

Lloyd, S. J., and Hensman, T. W. (1979). "The Pumping Characteristics of a Jet-Stream Feedback Fluidyne Pump." RNEC-SERG-4, April.

Malone, J. (1931). "A New Prime Mover." *J. Roy. Soc. Arts,* vol. LXXIX, no. 4099, pp. 697–709, June.

Martini, W. R. (1978). *Stirling Engine Design Manual.* 1st ed., NASA Rept. CR-135-382 (NTIS No. N78-23999), April.

Martini, W. R., Hauser, S. G., and Martini, M. W. (1977). "Experimental and Computational Evaluations of Isothermalized Stirling Engines." Pap. 779250, *Proc. 12th IECEC, Boston,* August.

Morash, R. T., and Marshall, O. W. (1974). "The Roesel Closed Cycle Heat Engine." Pap. 749154, *Proc. 9th IECEC, San Francisco,* August.

Mosby, D. C. (1978). "The Fluidyne Heat Engine." M.Sc. thesis, Naval Postgraduate School, Monterey, California, September.

Murphy, C. L. (1979). "Review of Liquid Piston Pumps and Their Operation with Solar Energy." ASME-79-Sol-4, pp. 2–8, March.

Pandey, R. B. (1981*a*). *Financial Express of New Delhi,* February 14.

Pandey, R. B. (1981*b*). Private communication to C. D. West, April.

Park, J. R. S., and Baird, M. H. I. (1970). "Transition Phenomena in an Oscillating Manometer." *Can. J. Chem. Eng.,* vol. 48, pp. 491–495, October.

Payne, P. R., Brown, R. G., and Brown, J. P. (1979). "Final Report: Water Pulsejet Research." Working Pap. 214-14, DOE Contract EG-77-C-01-4121, April.

Rallis, C. J., and Urieli, I. (1976). "Optimum Compression Ratios of Stirling Cycle Machines." Rept. 68, School of Mechanical Engineering, University of the Witwatersrand, June.

Rayleigh, Lord J. W. S. (1896). *The Theory of Sound,* vol. 2, 2d ed. Macmillan, London.

Reader, G. T. (1979). "The Fluidyne—A New Class of Heat Engine." Pap. 19, *Polytechnic Symp. Thermodyn. Heat Transfer, Leicester,* November.

Reader, G. T., Ivett, G., Gill, P., and Lewis, P. D. (1981). "Modelling the Jet-Stream Fluidyne." Pap. 819792, *Proc. 16th IECEC, Atlanta,* August.

Reader, G. T., and Lewis, P. D. (1979*a*). "Modes of Operation of a Jet-Stream Fluidyne." Pap. 799238, *Proc. 14th IECEC, Boston,* August.

Reader, G. T., and Lewis, P. D. (1979*b*). "The Fluidyne—A Water in Glass Heat Engine." J.N.S., vol. 5, no. 4, pp. 240–245.

Richardson, P. D. (1963). "Note: Comments on Viscous Damping in Oscillating Liquid Columns." *Int. J. Mech. Sci.,* vol. 5, pp. 415–418.

Rios, P. A., Smith, J. L. Jr., and Qvale, E. B. (1969). "An Analysis of the Stirling Cycle Refrigerator." *Advances in Cryogenic Engineering,* vol. 14, pp. 332–342, Plenum, New York.

Roesel, J. R., Jr. (1970). Engine. U.S. Patent 3,608,311 (filed April 1970).

Sandfort, J. F. Jr. (1964). *Heat Engines*. Heinemann, London.

Simonson, J. R. (1967). *An Introduction to Engineering Heat Transfer*. McGraw-Hill, New York.

Singleton, J. R. (1979). "The Fluid Mechanics of the Jet-Stream Fluidyne." RNEC-SERG-6-79, May.

Stammers, C. W. (1979). "The Operation of the Fluidyne Heat Engine at Low Differential Temperatures." *J. Sound Vib.*, vol. 63, no. 4, pp. 507–516.

Thwaites, G. C. (1979). "Development of Instrumentation to Enable an Experimental Analysis to Be Made of a Jet-Stream Feedback Fluidyne." RNEC-SERG-5-79, May.

Walker, G. (1973). *Stirling Cycle Machines*. Clarendon, London.

Walker, G. (1979). "Elementary Design Guidelines for Stirling Engines." Pap. 799230, *Proc. 14th IECEC, Boston*, August.

Walker, G. (1980) "Regenerative Engines with Dense Phase Working Fluids—The Malone Cycle" Pap. 809454 Proc. 15th IECEC, Seattle, August.

Walker, G. and Agbi, B. (1973a). "Thermodynamic Aspects of Stirling Engines with Two-Phase, Two-Component Working Fluids." *Can. Soc. Mech. Eng. Trans.*, vol. 2, no. 1, pp. 1–8.

Walker, G. and Agbi, B. (1973b). "Optimum Design Configuration for Stirling Engines with Two-Phase, Two-Component Working Fluids." ASME 73-WA/ DGP-1.

West, C. D. (1970a). "Hydraulic Heat Engines." AERE-R 6522, September.

West, C. D. (1970b). "Advantages of Reducing the Pressure Ratio in a Thermomechanical Generator." AERE-M 2351, September.

West, C. D. (1971). "The Fluidyne Heat Engine." AERE-R 6775, May.

West, C. D. (1974a). Improvements in or Related to Stirling Cycle Heat Engines. British Patent 1,487,332 (filed November 1974).

West, C. D. (1974b). Improvements in or Relating to Stirling Cycle Heat Engines. British Patent 1,507,678 (filed November 1974).

West, C. D. (1977). Fluidyne Development for Metal Box Overseas Ltd. Reading, England. Final Report. Private communication to Metal Box, March.

West, C. D. (1978). Private communication to the Metal Box Company, Reading, England. February.

West, C. D. (1979). ALPHA WEST. Westware Company, Rt. 3, Box 262A, Oliver Springs, 37840.

West, C. D. (1980). "An Analytical Solution for a Machine with an Adiabatic Cylinder." Pap. 809 453, *Proc. 15th IECEC, Seattle*, August.

West, C. D. (1981). "Theoretical Basis for the Beale Number." Pap. 819793, *Proc. 16th IECEC, Atlanta*, August.

West, C. D. (1982a). "The Stirling Engine with One Adiabatic Cylinder." ORNL/ TM-8022, March.

West, C. D. (1982b). "Performance Characteristics of Wet and Dry Fluidynes." in press, *Proc. 17th IECEC, Los Angeles*, August.

West, C. D., Cooke-Yarborough, E. H., and Geisow, J. C. H. (1970). Improvements in or Relating to Stirling Cycle Heat Engines. British Patent 1,329,567 (filed October 1970).

West, C. D., and Geisow, J. C. H. (1975). Improvements in or Relating to Stirling Cycle Heat Engines. British Patent 1,568,057 (filed November 1975).

West, C. D., Geisow, J. C. H., and Pandey, R. B. (1976). Improvements in or Relating to Stirling Cycle Heat Engines. British Patent 1,581,748 (filed April 1976).

West, C. D., Geisow, J. C. H., and Pandey, R. B. (1977). Improvements in or Relating to Stirling Cycle Heat Engines. British Patent 1,581,749 (filed January 1977).

West, C. D., and Pandey, R. B. (1981). "A Laboratory Prototype Fluidyne Water Pump." Pap. 819787, *Proc. 16th IECEC, Atlanta,* August.

Wood, G. (1980). Lecture notes for Stirling engine workshop. Sunpower Inc., 6 Byard St., Athens, Ohio 45701, October.

Name Index

Subject Index